MARS
3-D

A Rover's-Eye View
of the Red Planet

JIM BELL

STERLING

New York / London
www.sterlingpublishing.com

Half-title page: *Opportunity* Pancam 3–D image from sol 19. See pages 94 and 95 for more information.

Title page spread: *Viking Orbiter* mosaic view of thin clouds and hazes above Argyre Basin, in the heavily cratered southern highlands.

Photos courtesy of:
NASA/J. Bell: pages ii - iii
NASA/G. Legg: page 18
NASA/JPL: pages 3, 15, 23, 43, 53, 59, 71, 75, 115, 121, 125, 126
NASA/JPL/Cornell: pages ix, 9, 17, 27, 33, 38, 39, 40, 42, 44, 46, 47, 48, 50, 51, 52, 54, 56, 57, 58, 60, 61, 62, 64, 65, 66, 67, 68, 69, 70, 72, 73, 74, 76, 77, 78, 79, 80, 82, 83, 84, 85, 86, 87, 88, 89, 92, 93, 94, 96, 97, 98, 101, 102, 104, 105, 106, 107, 108, 110, 112, 113, 114, 116, 118, 119, 120, 123, 124, 127, 128, 129, 130, 131, 132, 133, 134, 136, 137, 138, 139, 140, 142, 142, 143, 144, 145
NASA/JPL/USGS: pages 31, 41, 45, 49, 55, 63, 81, 95, 99, 103, 109, 111, 117, 135

Library of Congress Cataloging-in-Publication Data

Bell, Jim.
  Mars 3-d : a rover's-eye view of the red planet / Jim Bell.
    p. cm.
  Includes index.
  ISBN 978-1-4027-5620-7
  1. Mars (Planet)--Exploration. 2. Mars (Planet)--Pictorial works. 3. Three-dimensional imaging in astronomy. I. Title.

QB641.B445 2008
523.43--dc22

                                                                2007049953

10   9   8   7   6   5   4   3   2   1

Published by Sterling Publishing Co., Inc.
387 Park Avenue South, New York, NY 10016
Text © 2008 by Jim Bell
Distributed in Canada by Sterling Publishing
c/o Canadian Manda Group, 165 Dufferin Street
Toronto, Ontario, Canada M6K 3H6
Distributed in the United Kingdom by GMC Distribution Services
Castle Place, 166 High Street, Lewes, East Sussex, England BN7 1XU
Distributed in Australia by Capricorn Link (Australia) Pty. Ltd.
P.O. Box 704, Windsor, NSW 2756, Australia

Book design and layout: Oxygen Design

Printed in China
All rights reserved

Sterling ISBN 978-1-4027-5620-7

For information about custom editions, special sales, premium and corporate purchases, please contact Sterling Special Sales Department at 800-805-5489 or specialsales@sterlingpublishing.com.

MARS
3-D

CONTENTS

Acknowledgments iv
Preface viii

ONE
Why 3-D? 1

TWO
Machine Vision 7

THREE
Can't Kill 'Em 13

FOUR
Daily Grind 21

FIVE
Not Your
Grandfather's Mars 29

Spirit Gallery 36

Opportunity Gallery 90

Web Links and
Other Resources 146

Index 147

ACKNOWLEDGMENTS

THE MARS ROVERS *Spirit* and *Opportunity* were conceived more than a decade before they were born, by a small group of scientists, engineers, and managers who had a remarkable group vision: that of being able to project themselves there through robotics, and then to explore the place remotely as if they were there in person. These colleagues, and the subsequent team of thousands of incredibly talented people from around the world who came together to carry out this vision, are the ones who deserve the biggest acknowledgment here: for bringing these rovers "to life" and for enabling me to share with everyone the remarkable photographic views showcased in this book. I am especially grateful to the engineers and technicians at NASA's Jet Propulsion Laboratory (JPL) in Pasadena, California, and elsewhere who designed and painstakingly assembled the Mars rover cameras to specific and exacting tolerances (enabling the high-quality, stereoscopic views depicted here). I would also like to thank the science and engineering team members who have been helping to write the sequences and to acquire the high-resolution color panoramic camera (Pancam), navigation camera (Navcam), hazard avoidance camera (Hazcam), and Microscopic Imager (MI) pictures from Mars for more than 2,500 sols (Martian solar days) combined as of this writing. We had hoped that at least one of the rovers would last 90 sols on Mars; our hopes were fulfilled, and indeed surpassed, in a most dramatic fashion!

Fusing left and right camera images into 3-D anaglyphs is a tricky business, and it depends greatly on the proper processing and geometric projection of the images. In that regard, I am very grateful to

Jonathan Joseph at Cornell University for developing the extremely capable and flexible Mermap and Merview image processing software, which allowed me to generate many of the 3-D as well as color views shown here, and to the team at the JPL Multimission Image Processing Laboratory (especially Bob Deen and Kris Capraro) for generating some of the anaglyph images using their own skills and equally excellent software. I am also very grateful to Ken Herkenhoff and his colleagues on the MI team at the U.S. Geological Survey in Flagstaff for suggesting and generating the 3-D views from the MI shown here, and to fellow Martian "travelers" Doug Ellison, Ken Kremer, and Nico Taelman for suggesting some of their favorite photos (so many photos to remember!). While these folks have been amazingly helpful on this project, I've done enough meddling on my own to assert that any shortcomings in the selection, processing, or visualization of the images are completely my own, not theirs.

On a personal level, I thank Michael Bourret from Dystel & Goderich for believing in this project from the beginning and (gently) helping to get it to the launch pad, and Meredith Hale and her team at Sterling Publishing Co., Inc. for expertly guiding it to a safe landing. And as always, I am grateful to my wife Maureen and my children Dustin and Erin for their support, patience, and indulgence while I was working to assemble this collection of wonderful Mars images and related stories.

JIM BELL

May 2008
Lansing, New York

PREFACE

I **LOVED 3-D PICTURES** as a kid. It was great fun putting on the silly red-and-blue glasses and then watching dinosaurs, bugs, and rockets jump out at you—like magic. One of my favorites was a 3-D ViewMaster cartridge that showed some of the early Mercury and Gemini astronauts and their spaceships—what a thrill! It felt like I was up in outer space with them.

For the past decade, I've had the incredible privilege of being in charge of the camera team for the Mars Exploration Rovers *Spirit* and *Opportunity*, the plucky, little wheeled robots that started driving around on Mars in early 2004 and which, as of this writing, are still going strong. Each rover carries four pairs of stereo cameras and a microscope that is also capable of stereo imaging. When we designed the cameras and started thinking about how they'd be used on Mars, we all knew that taking 3-D pictures would be an important part of their daily work. However, to be honest, I didn't think that we'd get that many 3-D images—certainly not enough to fill a book. In fact, most of us didn't think that the rovers would survive on Mars as long as they have. They were designed by the creative and skilled engineers at NASA's Jet Propulsion Laboratory (JPL) in Pasadena, California, to last for at least 90 Martian "sols" (a sol is a Martian day, about 24 hours and 39 minutes long), but they've survived—and thrived—for much, much longer than that. The result has

↺Rocky Hollows

Spirit acquired this 3-D Pancam mosaic while driving northeast across the volcanic plains toward the rim of Bonneville Crater. The impact that created Bonneville may have spread a large amount of previously buried material across the area.

been an enormous data set of images, more than 200,000 so far, including thousands of glorious 3-D views of an alien landscape at many different scales of magnification and from a variety of perspectives.

This happy circumstance has provided me with the chance to combine my childhood love of 3-D pictures with my grown-up day job as a Mars rover scientist and photography specialist. The results are compiled here in what are, in my opinion, some of the "greatest hits" 3-D images from both rovers during their nearly four-year adventure on Mars. About eighty 3-D photos are shown here, along with (wherever possible) a natural color or false-color view of the same or similar scene for comparison. Many of the 3-D images are striking: Rocks seem to pop up off the page, and sand dunes change from gentle, wavy landforms to menacing driving obstacles. Perhaps most impressive is the way the topography of the landscape, which often seems flat when viewed in a normal photo, jumps out at the viewer when seen in 3-D. Seen in three dimensions, the rovers' wheeled maneuvers across rolling plains, up rugged, rocky ridges, and down slippery impact-crater slopes are even more impressive. These images also provide a sense of the challenges that my colleagues on the team have faced in navigating these remote-controlled robots across the terrain, and those faced by the rovers themselves, when they've had to switch to their own self-navigation software.

For me, the most difficult part of viewing these images is the almost complete lack of any gauge that could be used to indicate the size of objects. There are no trees, shrubs, empty soda cans, or even geologist's rock hammers in these photos to provide a sense of relative size or distance. And since the rover itself is the photographer, there's often no way to get a wheel or some other part of the rover into the scene to make up for the lack of scale or context. Many of the rocks and sandy ripples that are close to the rover are much smaller than they seem; conversely, many of the hills and ridges that are seen off in the distance are actually much farther away than they seem. I've tried to indicate some sense of size and scale in the captions, where possible, but the images are still somewhat visually misleading at times.

You'll notice a lot of interesting or colorful names of rocks, craters, hills, and other features throughout the text and captions: Wooly Patch, Pot of Gold, Home Plate, Purgatory, El Dorado, etc. Giving names to new places and things is something that explorers have been doing for a long time. Most of the names that we use are theme-related to the place under observation at the time we're exploring the area. For example, rocks and other features on and near a broad plateau that we called Home Plate in Gusev Crater have been named after famous players from baseball's Negro Leagues and the All-American Girls Professional Baseball League. Sometimes we were more methodical, as when we named craters in Meridiani Planum after famous ships of exploration (*Victoria* was one of Magellan's ships; *Endurance* was the ship on Sir Ernest Shackleton's famous South Polar Expedition; and of course, *Eagle* was the ship that took Neil Armstrong and Buzz Aldrin to the Moon). Some of the names were suggested by school kids or members of the public who sent e-mails to various team members. Many times, though, names or themes just popped into teammates' heads on the spur of the moment—like rock-naming themes based on favorite ice-cream flavors, prairie wildflowers, or people and events from the French Revolution. None of these names are "official," though, in the sense of formally being recognized as feature names by NASA and the International Astronomical Union's planetary nomenclature committees.

The only officially named features have been those of the major peaks in the Columbia Hills, named to honor the astronauts who lost their lives in the Columbia Space Shuttle tragedy in 2003 (Commander Rick Husband and crewmates Michael P. Anderson, David M. Brown, Kalpana Chawla, Laurel Blair Salton Clark, William C. McCool, and Ilan Ramon), and three other prominent hills in Gusev Crater that are named in honor of the astronaut crew who lost their lives in the Apollo 1 fire in 1967 (Roger B. Chaffee, Virgil I. "Gus" Grissom, and Edward H. White). For everywhere else, naming things is just a fun activity that helps us to remember the places we've visited.

The main object of this book is to showcase some of the best 3-D photos from the rovers. Some of the accompanying text describes the ways we acquired the images and how we used them every day for scientific studies and for making decisions about how to drive the rovers or position the instruments. I've taken the liberty of expanding the text beyond just describing the 3-D imaging, however, in order to tell other stories about these exciting missions. Many of these stories are the expanded versions of answers to common questions that I get when I give talks around the country about the rovers: How do you drive a remote-controlled car from 100 million miles (160 million kilometers) away? Why didn't you install a wiper blade to clean the dust off the solar panels? How long are they going to keep going? Hopefully, these stories will help people understand some of the behind-the-scenes details about the kinds of decisions, skills, and experiences needed for a team of people to pull off an amazing feat like that of driving these rovers across the surface of Mars. And hopefully, the subjects of the 3-D photos shown here will jump off the page to amaze and delight you, just like the 3-D photos I viewed as a kid delighted me.

ONE

Why 3-D?

SOMEWHERE ON THE OTHER SIDE of the solar system, a lonely traveler is making its way across an ancient, dry lakebed, taking pictures and making measurements, every day, to share with all of us back here on planet Earth. This is no ordinary traveler, though: it is a robot, designed and built by earthbound humans to work and thrive independently on another planet. And this is no ordinary planet: it is Mars, a world, it would seem, that used to be a lot like the Earth but that has somehow dramatically transformed into the cold, dry, barren place that we know today.

The traveler—two travelers, actually—are the Mars rovers *Spirit* and *Opportunity*. They are the hardy and intrepid robotic explorers that were launched from Earth in 2003, which landed on Mars in 2004, and that have been cruising around the Red Planet for more than four years, making one scientific discovery after another.

A team of literally thousands of people—engineers specializing in robotics, rockets, power and communications systems, and navigation; scientists specializing in geology, astronomy, chemistry, physics, and even biology; supporting managers and administrators—designed, built, tested, launched, and then began to remotely operate these rovers on Mars from Earth. The team has had one main goal in mind through what has been more than ten years of planning and work on these missions: to use these amazing machines to project ourselves there in order to "virtually" explore Mars.

The rovers are extensions of ourselves and our very human capabilities. People can't yet go to Mars (though that day is coming), so we do the best we can given our limitations. The rovers were designed to be akin to mobile field geologists and, as such, possess many of the "senses" and abilities we would have if we could be out there exploring the place ourselves. We endowed them with mobility—the ability to move and drive and climb and descend. They move slowly, at a turtle-like pace, but with purpose and conviction, either relying on our judgment to help them navigate the remote terrain, or many times using their onboard software to make their own decisions about avoiding hazards or obstacles or otherwise finding the safest path to the next target of interest. We had hoped that the rovers could drive around—traverse—about 500 or 600 meters (about a third of a mile) at each of their landing sites. Instead, they've astonished us with their ability to scramble up rocky hills and descend down steep crater slopes, to the tune of more than 7,200 meters (more than 4.5 miles) of driving for *Spirit* and more than 11,500 meters (more than 7 miles) of driving for *Opportunity*. They've turned out to be marvelously mobile Martian machines.

We gave the rovers the ability to "feel" their environment using an agile, human-like arm that can reach out and touch the rocks and soils. The arm has four "fingers," made up of two instruments, that can measure the details of the chemistry and iron-bearing minerals on Mars; a small brushing and grinding device called the Rock Abrasion Tool (RAT), which can clean off dusty surfaces or grind small holes into rocks to allow their internal properties to be measured; and a Microscopic Imaging camera (or MI for short) for taking close-up pictures. The microscope takes pictures at about the same resolution as a geologist's hand lens. If we were sending robotic geologists into the field, we figured, they should be equipped with hand lenses. After all, we would be.

The rovers' cameras, their "eyes," are the final, most critical piece of this extension of ourselves into the alien Martian landscape. To a geologist, having mobility is good, as is being able to take samples and detailed measurements. But the most important thing geologists do in the field is *look* because, while the land is telling a story, preserved in the rocks and soils, it is a difficult one to read. There may have been

↻ Egress

This 3-D view of the *Spirit* rover rolling off its lander was generated by merging actual 3-D images of the lander on Mars with a 3-D computer model of the rover superimposed on the scene at the correct scale. The rover is about as tall as a typical ten-year-old kid.

violent eruptions of lava or floods of water. Or maybe there were glaciers, or ancient, long-ago buried rivers or sand dunes. The story is there, in the geologic record, but deciphering the clues requires careful observations and a fair bit of detective work. Looking for clues in the land's features is the key to untangling its past. It's the same for robotic geologists on Mars, which is why we knew we would have to give the rovers the best pairs of eyes that had ever been flown into outer space.

Each rover actually has nine "eyes." The MI is on the rover's arm; two fisheye-lens cameras are near the ground on the front and two more are on the back (these are the hazard avoidance cameras, or Hazcams); two wide-angle cameras are at the top of the mast (the navigation cameras or Navcams); and two high-resolution color panoramic cameras (Pancams) are also placed at the top of the mast for scientific observations. The Hazcams, Navcams, and MI are all black-and-white cameras; only the Pancams can take color pictures, by using thumbnail-sized filters to shoot, for example, red-, green-, and blue-filtered images that we can then combine later (once the photos are radioed back to Earth) into full-color (RGB) images. The resolution of the Pancams is about the same as that of a typical pair of human eyes on a person with 20/20 vision. However, some of Pancam's filters give the rovers superhuman color imaging capabilities by enabling us to see in the ultraviolet and infrared part of the spectrum through their eyes.

Having two eyes on a camera system enables the world to be viewed from two perspectives. The resulting parallax—just like the one that exists for our own eyes—is processed by the rover's brain (which is about as powerful as a mid-1990s PC) to yield a stereo view of distance and depth. Even images from the one-eyed microscope on the rover's arm can be used to "see" in stereo, by moving the arm to different positions in order to view the scene from a variety of angles. We can take the images viewed from the left and right sides and process them in a computer on Earth to simulate what the rover saw on Mars. The result, as you'll see in this book, is eye-popping. Rocks "jump" out of the photos. Sand dunes take on a feeling of waves cresting on the beach. Hills, ridges, and valleys suddenly appear where before there was only a flat world of two dimensions.

The Mars rover stereo images (also called "anaglyphs"—from the Greek word for "carved in low relief") can be stunning, even disorienting, to view, given their literally otherworldly nature. They are more than just interesting 3-D photos, though. During the mission, we used stereo images like these every day for practical, pragmatic purposes. For example, if we wanted to measure the size of a rock or a sand dune for scientific analysis, or assess the slopes and headings of different rock layers in the course of generating a geologic map, or accurately calculate the distance to a certain hill or ridge to triangulate our location, we'd use the parallax of the stereo images and the known resolution of the camera as a distance- and slope-finder. Or when we were planning to use the RAT to grind into a specific rock, first we'd take stereo pictures of the surface of the rock in order to build a 3-D computer model, and then the RAT engineers could use that model to choose the specific location and "angle of attack" for the grinding.

Every day, the rover drivers (believe it or not, there is a small team of "licensed" Mars rover drivers who work at the rover's mission operations center at JPL) use stereo images from the Hazcams, Navcams, and Pancams to build, inside a computer, a virtual-reality 3-D model of the terrain around the rover. They can then practice "driving" a computer model rover in this computer-generated facsimile of the real Martian world. The rover drivers have always worked closely with the scientists on the mission and the computerized rover/topography models to make sure that it was possible (and safe) to drive the rover to the places the scientists wanted to go. If things got dicey, they would then have to either find a way around obvious obstacles or over less steep slopes, or work with the scientists to choose a different target destination. Once the path forward looked good, the drivers would build the final "drive sequence" and program the real rover on Mars to carry out the same driving instructions that had been used to propel the computerized rover on Earth. Essentially, you can think of the rover's stereo images as part of an elaborate interplanetary video game being played every day.

We see and experience Mars through the eyes of the rovers. Spending time exploring the 3-D images taken by them is my favorite reward for having helped endow them with human-like vision. Our gift to them was vision and independence; their gift to us has been to show us the depth, texture, color, and beauty of the Mars many of us wish we could see one day with our very own eyes.

TWO

Machine Vision

THE ROVER'S STEREO CAMERAS and computer are modeled after the human eye and brain, which come equipped with one of the most incredible stereo imaging systems available. A pair of human eyes is a color, binocular stereo "camera," if you will, connected in a real-time feedback loop to a phenomenal central processing unit (our brain) that rapidly scans images, recognizes patterns, and performs necessary routine mechanical maintenance. Of course, our eyes and brains have evolved to be efficient at these tasks, many of which can be critical to survival (for example, Tiger approaching! Run!). The resulting combination of two high-resolution imaging "cameras" providing 3-D input to our superfast decision-making internal "computer" is the envy of and model for engineers, who then try to mimic these capabilities in robots and other electromechanical imaging systems. These researchers work in what is sometimes referred to as the "machine vision" field. There are strong links between this community and scientists and engineers working in the fields of computer programming, bioengineering, artificial intelligence, and, of course, space exploration.

Machine vision is just what it sounds like: the attempt to give machines the ability to "see" in a human sense and to make intelligent decisions based on what their sense of sight tells them. The potential of practical applications for successfully enabling machines to see and decide things on their own is enormous. One area in which this field has already taken off is manufacturing and process control. As products fly past on high-speed conveyor belts, for example, machines have been programmed with the ability to search for anomalies or defects (Is that top the right color? Is the package dented?) at a rate much faster than a human being could, and with fewer errors. There are a huge number of other applications for machine vision, like medical research and diagnosis, security and surveillance, transportation (navigation, traffic

7

monitoring, hazard avoidance), and remote sensing. Machines are reacting to what they observe all around us, and for the most part our society now takes their help for granted.

When those of us on the Mars rover design team started thinking about the kinds of capabilities the rovers would eventually be endowed with, machine vision was a critical part of the equation—especially for driving. It had to be. In essence, the rovers' sight had to "evolve" to be as efficient and useful as possible to their survival in the Martian environment. Perhaps the most environmentally relevant factor was that the rovers would be anywhere from about 60 million to more than 300 million kilometers (35 to 186 million miles) from their controllers back on Earth. Even at the speed of light, that meant radio signals carrying pictures and other data could take from about 5 to 20 minutes just to get from Mars to Earth, and it would take just as long or longer for instructions from home to get radioed back to Mars. By the time a hypothetical cry for help could be received from one of the rovers ("Hey, those commands you sent me are going to cause me to drive over the edge of that cliff. Should I stop?"), it might already be too late to send a reply. The alternative would be to send commands to make the rover do some minimal, safe activity and then to wait for images and other data to confirm that activity was not hazardous before going on to the next one. However, this method could be inefficient or impractical for some kinds of terrains or driving circumstances because each of these mini-decision cycles might take an entire day to complete (we're not able to be in constant communication with the rovers because NASA's Deep Space Network communications antennas need to support many other space missions, plus the rovers are out of view of the Earth for about half of each Martian day). Clearly, the rovers had to be able to make their own decisions sometimes, based on what they saw. And those decisions had

↺ Desolate Ridge

This 3-D view was taken by *Spirit*'s Pancam instrument. The goal was to get the rover to the horizontal ridge, seen halfway from the top in this view; stereo images like this one provided the team with the critical information needed to safely plan the drive.

to be the right ones, given that about $400 million of taxpayer funding had been invested in landing each rover safely on Mars!

While it may be essential in some cases, it's still not easy for scientists to turn over control of critical driving or data-gathering decisions to a robot running a computer program. The rovers are not "intelligent" *per se*; they are only as smart as their software, which in turn is only as good as its programmers. Even though their software was built and programmed by teams of very clever people, and was tested extensively, it can still be a difficult thing for people to be out of the loop, and trusting a robot to perform well in a hostile environment so far from home. For the most part (after a bit of on-the-job training and a few software tweaks), the rovers and their capable software designers have earned the trust of the entire science and engineering team. The truth is that we never could have traveled so far and sampled the geology, chemistry, and mineralogy of so many different kinds of places that we've visited if the rovers hadn't been provided with such great stereo cameras, coupled with autonomous decision-making capabilities.

I worry a little about the effect this positive experience might have on future missions and experiments, though, and whether it might place us on some kind of a scientifically slippery slope. How does a machine decide whether or not a picture contains interesting or uninteresting information? What factors go into choosing among multiple targets for more close-up examination? What kinds of measurements get priority if resources are tight? These kinds of questions have been debated extensively among our team and among people running space missions in the past, and the answers can't always be boiled down to a simple algorithm, especially for unanticipated discoveries or for more "qualitative" or context-dependent areas of research like geology. Sometimes, even in space exploration, people use their intuition—that is, they act on a hunch or roll the dice to see what comes up. How do we program those kinds of very human traits and propensities into a machine? Is there really such a thing as artificial intelligence?

These kinds of issues are actually part of the broader "humans versus robots" debate that has been going on for decades, and which is still going on right now within the space program. Which is better: a long-term, continuous program of relatively inexpensive but technologically and intellectually limited robotic missions, or more expensive and risky but less frequent planetary exploration conducted by humans? It's a tough call, and there are camps with firm opinions entrenched on either side of the debate.

It is easy to imagine the side of the debate on which most scientists who are heavily involved in robotic exploration missions like the Mars rovers come down on. However, my personal opinion on the "humans versus robots" question may be a bit unexpected: I'm a big supporter of human exploration. I believe that without a strong, vibrant, and destination-driven program of human space exploration, future successful robotic missions like the ones employing the rovers just aren't going to have the momentum to go forward. It was the *Apollo* and *Skylab* astronauts, after all, who first got me interested in and excited about space exploration and a possible career in planetary science. I think that kind of motivation is necessary at the national level as well. It is the human adventure of astronauts in space (including its dangerous, perilous, and tragic aspects) that excites and inspires the public and that helps to galvanize support in Congress and the Administration for continued space exploration funding—whether human or robotic.

My view is framed by more than just intangibles like inspiration, education, and national prestige, though. I think that people are going to be needed on other planets to use their judgment, experience, and understanding of context to truly promote new scientific discoveries. Robots have made possible a huge number of discoveries in our solar system; however, many of them occurred because it was the first time we had flown past, orbited, or landed on a particular place, so our robotic emissaries were (to borrow a phrase) "picking the low-hanging fruit" from our species' first phase of solar system exploration. We'll be finishing up that phase over the next several decades, however, which means that new discoveries are going to be more challenging to make: The remaining fruit is higher up on the tree. Robots are getting smarter all the time, but it's not clear whether robots' brains will ever be able to think on the fly or react to a hunch the way a human brain can. People—that is, astronauts with experience—may be required to tease out the most important discoveries of the future.

Sadly, the Space Shuttle and the International Space Station missions don't seem to have inspired or motivated the public the way the *Apollo* Moon missions of the late 1960s and early 1970s did. I think this makes the present an extremely risky and uncertain time for the future of space exploration. If NASA and other agencies can't get people—teachers, kids, space enthusiasts, and the general public—interested in space exploration, the entire enterprise (human exploration, robotic missions, and basic research) could wither on the vine. Part of the reason the Mars rovers have been so popular in the media and among the general public is precisely because they seem to offer the closest thing yet to having people on Mars. We can all move, virtually, from place to place there. We see Mars through their machine vision as we would see it through our own eyes. It's a destination we can imagine visiting, and there appears to be a nascent thirst among people to voyage there. Will we muster the will, the know-how, and the courage to make the trip ourselves? I hope so. The future of both human and robotic space exploration may depend on it.

THREE

Can't Kill 'Em

I VIVIDLY REMEMBER SITTING in a room listening to engineers talk about how the Mars rovers, *Spirit* and *Opportunity*, would die. The setting was an engineering presentation at JPL early in the history of the project. The room was full of scientists and engineers and managers, many of whom I'd never met but most of whom would go on to play critical roles in the success of the rover missions. The rovers had not yet been dubbed *Spirit* and *Opportunity*, just MER-A and MER-B, respectively. They weren't even born yet; in fact, they were still just twinkles in their designers' eyes. And yet, these people were already trying to figure out what would eventually kill their creations.

Cold, dry, barren, dusty, distant, and unforgiving: These words describe the environment the rovers would eventually be expected to call home. As scientists studying Mars—the most earthlike of all the other worlds in our solar system—we are sometimes guilty of glorifying what is, in reality, a truly desolate and inhospitable place. The pictures may at first look familiar, like a sunny desert scene, but with nary a cactus in sight. One expects to see a tumbleweed now and then, or a bleached jackrabbit skull. But Mars is nothing like a desert on the Earth. Far from it: The average temperature on the Red Planet is –50° to –60°C (–58° to –76°F)—close to the coldest temperatures ever recorded on Earth. On a balmy summer day in the Martian tropics, it might get up to 5° to 10°C (41° to 50°F) for a short while, but during a winter night near the poles, it drops down to –140°C (–220°F), an inconceivable cold during which the air itself (almost all of which is CO_2) freezes onto the ground. Even on the driest deserts of the Earth it rains, though perhaps only centimeters per century. As far as we can tell, it hasn't rained on Mars in perhaps 2 *billion* years or more. Talk about a drought!

So the rovers had to be built to survive the cold, with insulation and heaters to keep sensitive electronics warm, and batteries and solar panels to provide enough electricity to keep those heaters running at night. In fact, while I didn't appreciate it ahead of time, I've come to recognize that the single most important thing that the rovers do on Mars every day is recharge their batteries using electricity from the solar panels. If the batteries were to become drained so that we couldn't run the rovers' internal heaters at night, the temperature of the electronics, circuit boards, and other internal components would get so low that something would undoubtedly break. And that would be the end of the mission.

There were plenty of other ways the rovers could have died on Mars, or even before they got there. They could have blown up on the launch-pad in Florida if the Air Force/Boeing Delta II rockets that were used to launch them failed. They could have been shaken to pieces by the forces and vibrations of the launch itself. There could have been a problem with the upper stage of the rocket before the rovers left the Earth's orbit. Or a problem could have arisen with their cruise-stage power supply or other electronics during their seven-month flight to the Red Planet. Solar flares or galactic cosmic rays could have zapped the computers during the cruise as well. Indeed, that almost happened, as some of the largest solar flares in recorded history erupted during the rovers' trips to Mars in the Fall of 2003, bombarding both vehicles with high-energy radiation. Luckily, even this potential cause of death had been anticipated to a degree, and the built-in shielding to protect the rover computers and other electronics proved robust.

Then, of course, the rovers could have burned up in the Martian atmosphere if the entry trajectories were off by just a bit, or they could have crashed to the surface, shattering into a million pieces, if the parachutes failed to open or if any one of dozens of pyrotechnic devices failed to fire at just the right moment in their Rube Goldberg–style, airbag-assisted

ꕷ Escaping Eagle

Opportunity spent nearly two months studying Eagle Crater. This Navcam stereo mosaic was shot after the rover struggled to drive out of the crater. The rover finally escaped and was able to take this look back at its tracks leading back to its former Martian home.

landing systems. Or they could have landed safely, only to tumble to their deaths off the lander while trying to descend the final 35 centimeters (14 inches) needed to drive them down onto the surface. What a catastrophe that would have been—to come all that way and then to trip over the finish line and collapse! It was a failure scenario that someone was contemplating, though, which meant that there needed to be a plan to deal with it, and a backup plan in case that one didn't work. Even if we had landed them safely on the surface, with six wheels in the dirt, some unlikely software glitch or errant command sent from Earth could have shut off the transmitter or fouled up the software and effectively killed a rover. Indeed, this would have happened on *Spirit* just a few weeks after landing if not for some incredibly prudent contingency planning (even for extremely unlikely scenarios) by the rover engineers and software designers. Even the dryness of the Martian air could have killed them, potentially creating a lethal dose of static electricity as the rovers drove around. Care had to be taken to design circuits that were well-grounded and systems that were not prone to static discharges.

If the cold or the dry didn't kill them, though, then maybe the dust would. Martian dust is famous—it gives the planet its characteristic reddish color, and astronomers using telescopes have seen it coming and going for centuries, on occasion completely covering the planet for weeks on end. Mars dust is extremely tiny stuff. It's basically composed of smoke-sized, magnetic, and slightly rusted mineral particles only a few thousandths of a millimeter in size. Therefore, like soot and smoke on the Earth, it gets into everything. Wheels, gears, electronics, camera lenses—everything—will eventually get dusty and dirty, which could cause something to wear out or to jam or gum up. Most frightening was the threat posed by dust in the air (lifted by dust devils and dust storms) that it is always falling out and coating everything—like the rovers' solar panels.

↻ Bump in the Road

When the rovers are using their onboard software to "drive themselves," Hazcam images provide the information needed for the rover's automated onboard obstacle-avoidance software. This photo shows the front wheels driving backward over a 5 cm (2 inch)–high sand dune.

Viking Orbiter view of clouds around Olympus Mons, the highest volcano in the solar system.

Perhaps the favorite potential cause of rover death, often discussed, was the slow choking off of their electrical supply by dust settling onto the solar panels and slowly, inexorably, blocking out the sunlight. In fact, during the first few months after landing, we could see this happening in the daily power levels the rovers reported back to us. We were seeing two effects, actually, one of which was seasonal. Mars has seasons just like the Earth does, except that they are twice as long (the Martian year is 687 days). Both rovers landed in the midst of the southern hemisphere summer, and as summer made its transition into autumn, the Sun was getting lower in the sky every day, causing our power levels to steadily decline (the solar panels generate more electricity when the Sun shines directly on them, instead of at an angle). We had planned for that power-level drop, and we knew that if the rovers could survive until the winter solstice, the Sun would start getting higher in the sky again and their power levels would go back up.

The other effect we saw in our daily power levels was a gradual decrease in power as the solar panels got ever dustier. We could see this in the pictures, too: Our once shiny, gleaming blue rover deck was becoming a dusty, faded brownish red. The power kept dropping as the panels got dustier and the Sun got lower. Many of us could see the end coming, but no one really wanted to talk about it. We just kept pressing on, trying to get as much scientific exploring done with whatever time these amazing space vehicles had left to them.

And then, one day while *Spirit* was climbing to the summit of Husband Hill (the tallest peak in the Columbia Hills complex near the middle of Gusev Crater), a dust devil or a rogue gust of wind came along and—*swoosh*—cleaned most of the dust off the solar panels! All of a sudden our power levels jumped up. The deck was once again shiny blue, and we suddenly found ourselves with the power (literally) to do much more science and remote driving than we could do before. This happened multiple times, on both rovers, over the course of the mission. Certainly there was no way we could have anticipated the winds on Mars helping to keep the solar panels clean! As in life, sometimes in the space exploration business, you get lucky.

People often ask me why we didn't design some kind of wiper blade, compressed air sprayer, or other gadget to keep dust off the solar panels. It's a great question, and the answer may not be at all obvious. I tell them that it would have been relatively easy for the engineers to design something to clean the panels every so often—it's just a mechanical problem, after all, not rocket science. However (and here's the catch), there's only a fixed amount of mass, volume, power, and time available when you're designing, building, launching, and then operating a space mission. So adding some kind of panel wiper system would have meant getting rid of something else to offset the mass, volume, power, and time this new system would need. What would we remove? A camera? A spectrometer? A redundant radio transmitter? An extra battery heater? It's a tough choice because it's a zero-sum game. If we'd wanted to include a feature like that, we would have had to toss something else off. Our engineer friends assured us that the solar panels would give us the electricity we needed to keep the rovers going for at least 90 sols on Mars, even if they got dusty. So we decided to risk it and not do without any of the instruments or other systems in exchange. And they were right. The fact that we went beyond 90 sols—way, way beyond—was a fortunate bonus and a by-product of great design, careful planning, and yes, plain old dumb luck.

With spacecraft and space mission instruments, it is often said that things either break on their first day of use in their harsh new environment or they survive essentially forever. *Spirit* and *Opportunity* are testaments to that aphorism, and to the people at JPL and other government labs, universities, and hundreds of vendor companies around the world. They designed these amazing machines, provided parts and advice, wrote the software to run them, and tenderly guided them across the rocky, sandy plains far from their former home. Their designers came up with dozens of ways to kill the rovers, but after more than four years, Mars hasn't been able to come up with a single lethal blow. I think the rovers are like wild animals released into their native habitat, thriving in the environment they were built for, and truly fulfilling their birthright.

FOUR

Daily Grind

SOME DAYS I FEEL LIKE A TOURIST on a vicarious but virtual vacation to a far-off land. Those of us involved in the day-to-day operations of the Mars rovers come in to work in the morning and fill our computer screens with the latest pictures and other data from the Red Planet. Sand dunes, wispy clouds, frothy volcanic rocks, wheel tracks, and sometimes strange and alien landforms and features greet us daily. Many times it's the 3-D images, like the ones showcased here, that grab our attention the most. The strange part (as if our jobs aren't strange enough already!) is that many of us have gotten used to it. In fact, I don't really even remember what my day job was like before the rovers got to Mars. (What did I do with all that free time?) We've come closer than any other people ever have to *being* on Mars and experiencing the place as if we were really there.

When the rovers landed on Mars, all of us earthbound scientists, engineers, and rover drivers had to live our lives on "Mars time." That is, we had to react to and operate by the cycle of sunrises and sunsets that the rovers experience on Mars, because sunlight is what provides the power that the rovers need to work. However, a Mars day is about forty minutes longer than an Earth day, so our workdays here on Earth—and our lives in general—were slowly shifting by an increase of forty minutes each day, because the Sun rose on each rover forty minutes later each Earth day. It was fun for a while, working crazy hours and feeling detached, in a way, from our home planet even though we'd never actually left. I think it helped build a bond between us and our robotic friends on Mars. "Who's controlling whom here?" we wondered, while trudging in to start our shifts at 3:00 A.M.

During those first three to four months, several hundred of us lived and breathed Mars and rovers at JPL in Pasadena. The rovers had landed near the Martian equator (lots of sunlight) but on opposite sides of the planet: *Spirit* in Gusev Crater, a possible ancient lakebed the size of Connecticut, and *Opportunity* in Meridiani Planum, a flat, sandy plain the size of the western United States where evidence of water-formed minerals had been detected from orbit. As the Sun set on one of the rovers, the rover would radio its latest pictures and other telemetry information to us and then proceed to "sleep" for the night, hunkering down with its heaters to keep its systems warm and alive until the next morning. At the same time, on the other side of the planet, the Sun was just rising on the other rover, "waking it up" to begin its workday.

Our attention first focused on looking at the pictures and other measurements just sent down to Earth by whichever rover was then going to sleep. The clock was ticking on us all the time: We had to have a new set of instructions ready to transmit to that rover as soon as it woke up the next morning. That meant that we had to process the images and do some very quick analysis and assessment work to help make informed decisions about what to do, usually within just a few hours of when the data arrived. Should we tell the rover to drive? If so, which way? Or should we put the arm down on some targets to make detailed measurements of the chemistry and minerals, or photograph some big panoramas? While we always had well-thought-out strategic goals that we were aiming to meet, every sol presented us with new tactical challenges and opportunities to address. It was grueling but exciting work.

Once the constraints on the next sol's activities were understood (power, data volume, etc.) and the high-level tactical scientific decisions had been made (whether to drive, deploy the arm, take remote sensing camera/spectrometer observations, etc.), the part of the science and engineering team that concentrates on downlink—the assessment and analysis of data sent "down" to us from Mars—had essentially finished the most stressful part of its workday. Sometimes there would

↷ Mazatzal

This is a colorized *Spirit* 3-D Navcam mosaic showing the area near the rock Mazatzal. About 2 meters (6.5 feet) across, Mazatzal is the bright rock on the left of this view. A Rock Abrasion Tool (RAT) hole ground into the rock revealed it to be made of dark volcanic minerals.

be time to look at the data more in depth, or to more thoroughly investigate some scientific or rover system mystery that appeared in the data, before the next downlink would arrive. Sometimes, though, the downlink assessment and tactical planning process would have been so lengthy and stressful that once over, all you would have wanted to do was lie down on a cot and sleep.

The action would then continue on to the "uplink" process—the act of building the detailed instructions and commands to send up to the rovers on Mars. The clock was still ticking—we might have eight hours or so left until the rover was going to wake up. The uplink team was generally made up of different people from those on the downlink team. This was done partly for practical reasons, such as trying to keep people's workday to shifts under twelve hours. But partly it also reflected the fact that sending commands to run rovers and other space instruments is often much more of an engineering and software detail-oriented activity than is science or telemetry downlink assessment. People working on the uplink shift listen to the downlink reports and follow the tactical decision-making process; then they have to be the ones to implement the results of those decisions—in detail. For example, it's one thing to agree that you're going to take a panorama but another thing entirely to actually write the computer commands that point the camera, turn the filter wheels, take the exposures, etc., and then string together all the commands into detailed sequences that will be radioed to Mars.

Uplink people tend to be detail-oriented—careful, methodical, double-checker types. They are the kind of people who can spot a typo within seconds on a page full of words. They also tend to be extremely creative problem-solvers, and many of them are excellent at visualizing the situations they're writing sequences for and doing the 3-D math (often in their heads) needed to understand the detailed pointing of the instruments. They are exactly the kind of people you *want* to have driving (literally or figuratively) a rover or an instrument like a camera, because they rarely make mistakes. I'll confess that I'm not an uplink person—I don't have the patience for the level of detail and scrutiny their job requires, and I can't do 3-D vector math in my head (or on paper, for that matter). I'm much happier crunching the numbers and working with the pictures and other data on the downlink end of the process. It's all a matter of personal preference and individual skill and training. There is variety enough in the kinds of jobs required to be done to make for a team of people with extremely diverse interests and capabilities.

The uplink process results in the creation of a long list of sequences that the rover will start to run once it wakes up in a few hours. The list is made to fill up every available minute of time and every available watt of power. It's essentially a "to do" list for the rover, with a bunch of time tags telling it when to do each activity. For example, "10:53 A.M.: Begin battery recharging cycle; 11:20 A.M.: Begin drive 10 meters forward; 12:00 P.M.: turn 20 degrees left; 12:50 P.M.: drive 10 more meters forward; 1:30 P.M.: acquire 360 degree panorama," etc. The list is laboriously assembled, sequence by sequence, and is then checked and double-checked in the remaining hours before being transmitted. Sometimes the team will need to try out sequences on the test rover or simulator at JPL before sending them to the real rover on Mars. Once everyone's happy with the list, it's converted into the ones and zeros of spacecraft radio signals and transmitted by one of the NASA Deep Space Network telescopes, directly to the newly awakened rover. Then, we're out of the loop—the rover goes off and does everything on the list autonomously, and we usually don't hear back from it until the end of its day, when it radios back the results from that day's activities.

Meanwhile, around the same time the list described above is finally radioed up to the rover, the rover on the other side of the planet (remember, there are two) would have just finished working through the list that we sent it yesterday, radioed its results back to Earth, and be getting ready to sleep. A second subset of the rover team would then begin working through the downlink assessment, tactical planning process, and finally the uplink sequencing process for that rover, meeting the deadline of getting its new list uplinked before it woke up the next morning. When that rover would wake up and get its new to-do list, the other rover would be radioing back its results, in keeping with its daily agenda, and so on. One of them was always working during the daytime, and we were always working on planning the next sol for whichever one was currently sleeping. Essentially, the Sun never set on our little Martian empire! The only ones getting any decent sleep were the rovers....

The 3-D images from the Hazcams, Navcams, and Pancams were a critical part of this decision-making process every sol. Drives could occur only if the rover operators understood the topography and potential obstacles the rovers might face and felt confident that the rovers could negotiate parts of the terrain on their own, if necessary. The arm could be positioned and its instruments deployed onto rocks or other targets only if the shapes of those targets and the nature of the "work volume" where the arm could move was fully understood. Many times, we were basing our choice of scientific targets on detailed 3-D stratigraphy and on the geologic relationships between layers of rocks or other features that we could assess using the stereo vision capabilities of the cameras.

Working on this crazy cycle of round-the-clock care and feeding of robots on another planet was certainly a unique experience, but shifting our jobs 40 minutes later each day turned out to be so foreign to our natural circadian rhythms that it began taking a toll on our bodies and minds. By the end of the rovers' 90-sol primary mission on Mars, people were tired, irritable, and yearning for the good ol' 23- hour 56-minute day that we were used to back on Earth. So the mission managers brought us all back to our home planet, sent most of the scientists back to their home universities and labs, and switched us to an "Earth time" schedule for the rest of the mission. It was a little sad to separate ourselves from the daily cycles of the rovers and from the nerve center of the operations at JPL, but there really wasn't any other choice. It's just not possible to live on two planets at once, especially if you have a real life on one of them.

↻ The West Spur

This Pancam 3-D mosaic shows the slopes and ridges of the West Spur of the Columbia Hills, a region *Spirit* arrived at after nearly five months and more than 3.2 kilometers (2 miles) of driving.

The vast majority of the mission has since been run on Earth time instead of Mars time, and the team, now distributed around the world and networked through telephones and the Internet, has become extremely efficient and streamlined. We still go through the daily down-link and uplink cycles, but the decision-making and sequence-making processes generally take much less time than they did before. To save money (and perhaps some marriages), the managers eventually decided to give the team weekends off. However, the rovers don't get a break: On Fridays we plan each rover's activities for Saturday, Sunday, and Monday. The mismatch between Earth time and Mars time also means that occasionally we have to "skip" an Earth day planning cycle and instead plan two Mars days in a row. It sounds complicated, but we've actually gotten into quite a workable groove. Running the mission on Earth time and using so-called "distributed operations" has turned out to be a successful and efficient way to explore Mars remotely, and it has allowed team members to share more deeply the amazing results from these rovers with their families, neighbors, students, and broader local communities around the world.

FIVE

Not Your Grandfather's Mars

SPIRIT **AND** *OPPORTUNITY*'**S** discoveries have fundamentally changed our view of Mars. The way that science operates best is by people making observations and developing models of those observations in order to pose hypotheses that can be tested. In the history of Mars exploration, several decades of previous spacecraft observations had revealed vast channels and valley networks carved into the rock, huge past rates of erosion, deposits of possibly water-formed minerals, and even some geologic evidence for rainfall, lakes, and glaciers. These observations led to the hypothesis that Mars used to have liquid water on its surface, perhaps for a significant span of time early in the planet's history. The idea led to the further hypothesis that such a warmer and wetter Mars may have been a *habitable* and much more earthlike world. These hypotheses are testable, so verifying them by studying the geology, geochemistry, and mineralogy of small, specific regions on the surface was the primary goal of the rovers and their instrument suites.

One of the last decade's most profound discoveries about life on our own planet is that it is dominated by simple, single-celled organisms, and that these organisms (with which all more complex organisms ultimately share common ancestors) occupy and thrive in an incredibly wide range of niches and environments. From acidic pools to arid deserts, from the bottom of the seabed to the tops of the world's coldest glacial peaks, life abounds on Earth.

What makes a world habitable? This has become a central question for astronomers, planetary scientists, and biologists studying Mars and other heavenly bodies, as well as for geologists, biologists, paleontologists, and planetary scientists studying the Earth itself. Life is found essentially anywhere on Earth where it can avail itself of three main ingredients: water, energy, and organic molecules. Earth is a water world, three-fourths covered by salty oceans, harboring a global network of surface and subsurface freshwater, and still actively leaking out water and other volatile gases from the deep interior through volcanoes. Our planet has been described as being in a perfect "Goldilocks" place for water in our solar system: Not too hot (close to the Sun) to allow the water to boil off as on Venus, and not too cold (far from the Sun) to keep the water in a deep freeze like on Mars today. In other words, just right.

The Sun provides the main source of energy that enables life on our planet, through almost magical processes like photosynthesis, which converts sunlight into energy that can be used by cells (photosynthesis is to our cells what the solar panels are to the rovers). There are other ways to harvest energy, however. For example, bacteria deep underground extract energy from the rocks by slowly oxidizing them. Exotic life forms at deep mid-ocean ridges extract energy from volcanic heat and sulfur-rich gases and mineral deposits. Sunlight is an abundant energy source, but it's not the only source required for life—especially life underground or underwater—to thrive.

Earth is also endowed with a rich supply of organic molecules—the building blocks of life as we know it (there have been serious studies about "life as we don't know it," but by definition that is extremely hard to search for and identify; still, people are looking into it...). The elements carbon, hydrogen, nitrogen, oxygen, phosphorus, and sulfur can combine in ways that are simple or phenomenally complex to form the familiar organic molecules like proteins, carbohydrates, amino acids, and ultimately RNA and DNA, which constitute the basis of life on Earth. The Earth was formed with some of these chemical elements "built in," but others have been "delivered" to the Earth by comet and asteroid impacts throughout time.

↻ Wishstone

This *Spirit* Microscopic Imager 3-D mosaic shows part of a RAT hole ground into the rock called Wishstone. The rock's inner surface shows that it is composed of fragments of many different sizes and textures, possibly indicating that the fragments were fused together by some violent geologic event.

Might other planets, Mars in particular, have had or still have these three main ingredients? And might they exist or have existed in the right proportions to enable these places to be habitable? Mars has ample sunlight available (though only about half as much per square meter as the Earth). It also appears to have an inventory of rocks and minerals that could provide alternative energy sources similar to the ones exploited by organisms living deep inside the Earth.

Does Mars have water? Absolutely—we've known for decades that there is water ice in the polar caps and a tiny amount of water vapor in the atmosphere. Recent space missions have also discovered large deposits of underground ice at high to middle northern and southern latitudes. However, life (again, as we know it) appears to require *liquid* water, as a medium for chemical reactions and as a way to enable the mobility needed to bring in nutrients and to take out waste. So, does Mars have liquid water? Not today, at least not on the surface. The temperatures and pressures are too low for liquid water to be stable at or very near the surface (it would quickly "boil" off into water vapor). However, evidence from orbital data and now from rover data acquired at two specific sites indicates that Mars did have liquid water on and/or very near its surface long ago, when the planet was young. Minerals have been found that can only be formed in liquid water; some of them still have the H_2O or OH molecules derived from that water in their mineral structures. Geologic features, like preserved shallow water-wave ripples and the iron-rich spherical grains known as "blueberries" at the *Opportunity* rover site (which appear to have precipitated out of ground-water like stalactites or stalagmites in a cave), also provide evidence of water at one time flowing over the surface and saturating the rocks. The evidence seems incontrovertible: Mars used to be warmer and wetter. Its environment used to be a lot more like the Earth's.

↻ The Road Ahead
The boulder-strewn slopes and ridges seen here in this *Spirit* Pancam 3-D mosaic represent an enormous set of potential driving obstacles to the rover, which nonetheless had to be guided through such terrain to try to reach the summit. Happily, the rover turned out to be a great hill climber.

Does Mars have organic molecules? This is a more difficult question to answer. Experiments on the Viking Lander missions in the 1970s were designed to answer this question, and the results are widely interpreted to have been resoundingly negative. However, those missions literally only scratched the surface, and the landing sites didn't turn out to be particularly interesting or perhaps even very "water-related," compared with the rover sites or many other potential landing sites on Mars. Like Earth, throughout its history Mars has been bombarded by organic compound–bearing comets and asteroids, and Mars was almost certainly endowed with a built-in inventory of organic molecules, or at least their precursor elements. So the answer is potentially, "yes," though we lack the slam-dunk proof that we have for other parts of the habitability question. Indeed, this is exactly why the next two missions planned for Mars—the 2008 Phoenix Mars Lander and 2010 Mars Science Laboratory (MSL) rover—will be carrying elaborate organic chemistry instruments to their landing sites. *Phoenix* will land near the north pole of Mars and dig deep into the subsurface to search for ice and underground organic molecules. *MSL* will land in a place chosen for its geologic and/or mineralogical evidence of past surface water, and then will conduct detailed organic chemistry experiments on the materials exposed there. The results from these next few missions could be extremely enlightening and could help fill an important gap in our understanding of the habitability of Mars.

Still, the rovers have helped us to make a giant leap forward in our view of the Red Planet. To my grandparents' generation, Mars was just a reddish point of light in the sky and the topic of often wild, sensational speculations about its possibly nefarious inhabitants. It's not our grandparents' Mars anymore, however. We've been there and have traversed (albeit virtually) its ruddy surface. No longer do we have to speculate about whether Mars has potentially habitable, earthlike environments—we now know that it once did. We searched for the "smoking gun"—evidence of the planet's watery past—and we found that evidence, either lurking there in the rocks and minerals or popping out at us as soon as we looked up close, much like 3-D images on a printed page.

The rovers have provided new and important information about two places on Mars, but by working with instruments on many different orbiters high overhead, the story's scope can be expanded to include the entire planet. What other environments there might be habitable? Where should we send the next Mars missions to maximize the chances of answering that question? There's also a burning question that the rovers haven't been able to address: How long was Mars earthlike? If it was a mere flash-in-the-pan early in the planet's history, then there may not have been enough time for life forms to gain a foothold. However, if Mars remained earthlike for an era of "real" geologic time— a billion years, maybe more—then we might be able to make a case here in our own backyard for parallel planetary development.

How exciting and meaningful it is to realize that we're actively trying to answer one of the most fundamental, far-reaching questions we can ask as human beings: Are we alone? Even if we were to discover only simple forms of life on another planet such as Mars, the implications would still be profound. It would mean that there might be other habitable environments right here in our own solar system, and that more broadly, the universe is potentially teeming with life. Maybe one day we will be able to reach across the stars and make contact with our interplanetary neighbors. For now, it's thrilling enough to be taking baby steps in that direction from the surface of Mars.

Bonneville crater

sol 50

sol 100

Gusev Crater Plains

Landing Site
sol 1 Jan. 4, 2004

sol 125

Gusev Crater Plains

Spirit rover traverse path

0

500 m

0

0.3 miles

SPIRIT

Landed January 4, 2004

Gusev Crater

Columbia
Hills

North

sol 150

sol 300

sol 400

sol 200

sol 500

West Spur

sol 600

Husband
Hill

sol 700

El Dorado

Southern Basin

sol 750

Home Plate
(sol 750-1400 +)

Six Wheels on the Ground

↩ **Opposite:** Twelve Martian days after landing (sol 12), *Spirit* rolled down one of the lander's cloth ramps and finally touched down on the surface of Mars. The 3-D image opposite, merged from the left and right rear hazard avoidance camera (Hazcam), shows the results of the rover's successful egress onto the dusty soil. The rover's rear wheels, each about 25 cm (10 inches) tall, can be seen in the lower foreground.

Above: Pancam color photo of the *Spirit* lander taken a few sols after egress (sol 16). The deflated airbags that surrounded the lander during the landing on Mars, once pristine and white, are now reddish and dirty after having bounced around so much on the dusty surface. Hundreds of people helped design and build the lander, and seeing it there on Mars was a source of great pride to everyone involved.

Dust in the Wind

↺ **Opposite:** *Spirit* sol 15 Microscopic Imager 3-D view of a tiny 3 cm (1.2 inch)–wide patch of fluffy, clumpy Martian dust near the lander. Individual dust grains are microns wide (about the size of smoke particles) and thus too small to resolve with the microscope. But clumps and mini-towers of aggregated dust grains can be made out in the image. This kind

Above: *Spirit* sol 12 Pancam color image. The area shown here is a typical piece of Martian ground about 30 cm (1 foot) wide. Volcanic rocks are scattered about the scene. The 3-D dust image opposite was taken in the smooth patch just to the right of the center of this image.

Adirondack

↺ **Opposite:** This is a *Spirit* rover Front Hazcam 3-D view, taken on sol 15, of the football-sized rock named Adirondack. The rover took several days to position itself such that the rock was right between the front wheels, in the "work volume"—the perfect spot for the rover's arm to reach out and detect the rock's chemistry and other properties. Rover drivers need 3-D views like this to generate a computer model of the terrain and ensure that the rover can place its instruments on a target safely.

Above: Adirondack was also photographed in color by Pancam (on sol 14) in order to identify the best place to brush and grind it with the RAT. From a study of these kinds of pictures, it became clear that the team would need to use the RAT to grind through the rock's dust coating to determine the intrinsic character of the rock itself.

Daily Grind

↩ **Opposite:** *Spirit* Microscopic Imager 3-D picture of part of a 4.5 cm (1.8 inch)–wide, 3 mm (0.1 inch)–deep RAT hole ground into the rock Adirondack. The grinding tool created a smooth, polished surface and allowed small crystals and veins to be identified inside the rock. By monitoring how much energy the RAT used to grind into Adirondack, scientists could ascertain that it is as hard as typical volcanic rocks found on the Earth.

Above: This *Spirit* Pancam color image of the Adirondack RAT hole from sol 35 is a dramatic demonstration of how dusty the rock really is. Adirondack is actually a very dark, almost black rock. As the RAT ground below the dust and into the rock, it generated fine black powder that was then blown across the rock by the wind. Analysis of the material inside Adirondack revealed that it is, indeed, made of pristine volcanic minerals.

Shores of Bonneville

↺ **Opposite:** *Spirit* took this 3-D Pancam panorama of 200 meter (650 foot)–wide Bonneville Crater when it arrived at the rim on sol 67. The rover had to scramble over large boulders and rugged terrain to reach this dramatic vista point. Barely visible on the far left rim of the crater are the crashed remains of the heat shield that had protected the rover during its fiery entry into the Martian atmosphere.

Above: This is part of a large Pancam color panorama of Bonneville Crater acquired on sols 68 and 69. The interior and floor of the crater exhibit dunes and wind streaks but no evidence of the layering or color variations that might indicate different rock types with depth. Thus, despite the spectacular view, exploring Bonneville ended up being some-what disappointing because *Spirit* never found the ancient buried lake deposits scientists were hoping to discover there.

Dusty Dunes

↪ **Opposite:** On sol 73, *Spirit* acquired this 3-D Microscopic Imager photo of a 3 cm (1.2 inch)–wide piece of a small, dusty sand dune on the Gusev Crater plains. The rover had dug into part of the dune with its wheel in order to study the materials inside. The sand grains seen here are from the inside of the dune, and their sizes and shapes indicate that the wind was very active in its formation. This originally black-and-white view has been colorized to simulate the approximate natural color of the scene.

Above: *Spirit* Pancam sol 73 color photo of the "scuff mark" created by the rover wheel in the dune. Dunes like this have been seen throughout *Spirit's* journey through Gusev Crater, and orbital photos of other places on Mars indicate that dunes are quite common across the planet.

Boulder Ridge

↶ **Opposite:** This is a *Spirit* Pancam 3-D image of a rock-covered ridge on the Gusev plains drive from Bonneville Crater toward the Columbia Hills (sol 87). Some of the boulders seen here are 1 to 2 meters (3 to 7 feet) across, presenting significant challenges to driving the rover. Fortunately, stereo images provided the critical information needed for navigating a path around such obstacles. This originally black-and-white view has been colorized to simulate the approximate natural color of the scene.

Above: This sol 87 *Spirit* Pancam false color image shows the same boulder-covered ridge. Brighter, probably dust-coated rocks are mixed with darker, less dusty rocks in this jumbled terrain. Some of the rocks may have been thrown here as "ejecta" from the impact that created nearby Bonneville Crater, and others may be ejecta from more distant impact events. Still, all of them appear to be bone-dry volcanic rocks.

Hollowed Ground

↩ **Opposite:** *Spirit's* shadow makes for a nice self-portrait in this sol 111 Navcam 3-D image of sand and rocks near a small hollow in the Gusev plains. Stereo imaging of these small circular depressions helped show that they are heavily eroded and partially filled impact craters.

Above: Color views of the hollows, like this Pancam mosaic from sol 111, show them to be filled with very fine-grained reddish sand and dust. Apparently, as fine grains move across the surface or settle out of the air from dust storms, they become trapped within small craters and slowly build up over time, creating the relatively flat, bright, dusty hollows.

Pot of Gold

↶ **Opposite:** Distant images suggested that the geology might be dramatically different in the Columbia Hills. This hope was vindicated by close-up 3-D images like these from *Spirit*'s Microscopic Imager on sols 163 and 170. Here is one of the first radically different kinds of rocks studied by the rover: Pot of Gold (so named because it was "at the end of the rainbow" after the rover's nearly five-month-long drive to the Hills). The surface has small nodules and protrusions, suggesting that substantial physical and possibly chemical weathering has occurred.

Above: This *Spirit* Pancam false-color view of Pot of Gold, taken on sol 167, reveals the strange nodular appearance of this potato-sized rock and highlights the less oxidized (bluish in this rendering) qualities of many other nearby rocks and rocky fragments.

Wooly Patch

⤹ **Opposite:** Another dramatic find in the Columbia Hills was that the rocks there are much softer than the hard volcanic rocks of the plains. This can be seen in these *Spirit* Pancam sol 200 views of twin RAT holes ground into the rock called Wooly Patch. Each hole is about 4.5 cm (1.8 inches) wide and 5.2 mm (0.2 inches) deep. This view has been colorized to simulate the approximate natural color of the scene.

Above: This *Spirit* Pancam color view of Wooly Patch reveals that the color differences between materials inside the rock and those on the surface are not as dramatic as those seen in the Gusev Crater plains (for example, in the RAT hole shown previously from the rock Adirondack). This kind of information supports the idea that these rocks were weathered or altered in some way, probably by water, long ago in Mars' history.

Longhorn

↺ **Opposite:** This 3-D *Spirit* Navcam mosaic of the outcrop region known as Longhorn, near the summit of the West Spur in the Columbia Hills, was shot on sol 203. As the rover continued to climb up the slopes of the West Spur, the team was rewarded with spectacular new vistas nearly every day. This was the most difficult driving yet undertaken by the rover, and stereo images like this were critical for mapping out the safest possible routes by the rover drivers.

Above: This *Spirit* Pancam four-image mosaic of the Longhorn outcrop was taken on sol 210. The atmospheric clarity enjoyed at this point in the mission helped give the scene a spectacular sense of depth: The plains fade off into the distance to reveal the rim of Gusev Crater itself, nearly 80 kilometers (50 miles) away.

Flanks

↺ **Opposite:** *Spirit*'s ascent to the top of the West Spur was only a prelude to even more climbing, with the goal of reaching the tallest peak around—Husband Hill. 3-D mosaics of the flanks of Husband Hill, like this one from sols 229 and 230, revealed new evidence to account for complex layers and rugged ridges.

Above: Pancam color mosaic of the flanks of Husband Hill, along the traverse on the West Spur. Some of the layers and ridges, seen in the 3-D image, show slightly different colors in this view, suggesting that there might be a change in the chemistry or mineralogy in those areas. Using colors to track layered rocks across large regions is a common technique used by geologists on Earth. The rovers have shown that it can be a useful technique on Mars as well.

First Impressions

↺ **Opposite:** This *Spirit* sol 240 Microscopic Imager 3-D photo of fluffy, dusty soils shows a donut-shaped indentation mark made by pushing into the dust a 3 cm (1.2 inch)–wide contact plate on one of the rover arm's spectroscopy instruments. Tests like these are sometimes done to try to determine how strong the surface is and how fine (or coarse) the particles are. Here the dust grains are so fine that they compress easily, even under light pressure.

Above: This sol 263 Pancam color view shows the location of the accompanying 3-D image, for context. The compressed region is within a small "drift" of dust that has gathered between several rocks. Many scientists assume Martian dust is the same every-where, because the wind mixes it all around. The team periodically tests this assumption by measuring the chemistry and properties of dust piles like this, which are found at different places along the rover's course.

What a Drag

↺ **Opposite:** This *Spirit* Navcam 3-D mosaic, looking behind the rover at wheel tracks imprinted in the soft sand, was taken on sol 313 during the rover's traverse from the West Spur to the flanks of Husband Hill. Even though the terrain here is only gently sloped, loose, sandy soil like this still represents a significant potential driving hazard for the rovers.

Above: This Pancam color view is part of the "Thanksgiving Panorama" taken along the drive to the flanks of Husband Hill from sols 321 to 325, in late November/early December 2004. *Spirit*'s tracks can be followed back more than 100 meters (330 feet) into the distance. What was supposed to have been several days of straight-line driving turned out to look more like a drunken sailor's swagger through these soft sandy soils. Still, despite the slipping and sliding, the rover's software was usually able to compensate and keep *Spirit* on the right path.

Slip Sliding Away

↺ **Opposite:** This *Spirit* Navcam sol 338 3-D mosaic and Rear Hazcam sol 345 3-D image (inset) reveal evidence of some difficult driving along the slopes of Husband Hill. The tracks on the left represent the rover's initial attempt to drive toward the distant ridge. However, *Spirit* began slipping and getting bogged down in the soft sand encountered there. The team decided to turn the rover around and high-tail it out of this hazardous terrain.

Above: This *Spirit* Pancam sol 351 color image zooms in on some of the rocks and soils the rover dislodged. In a bit of serendipity, the rover accidentally dug up some brighter, bluer soil deposits that had been buried under the dusty uppermost layer. These soils turned out to have a higher abundance of salty sulfur-bearing minerals—one of the first indications that there once may have been significant water-related processes at work in this part of Gusev Crater.

Craggy Ridge

↺ **Opposite:** Challenging drives along and across steep rocky ridges were a continual theme in *Spirit*'s ascent of Husband Hill, which took nearly one Earth year to complete. A dramatic example is seen here in this sol 391 Pancam 3-D mosaic of rocks and craggy ridges approaching the summit. Stereo images like this were used to assess the slopes and determine whether the rover and drivers would need to find another way around.

Above: This *Spirit* Pancam sol 392 false color photo shows rocks along the crest of the ridge seen in the accompanying 3-D view. Many of the rocks here are dusty and wind-scoured, but some of them, like the one in the fore-ground, appear much "fresher" and less dusty. This rock and others like it may be relatively younger pieces of debris launched here from distant volcanoes or impact craters.

Whale of an Outcrop

↺ **Opposite:** This *Spirit* Front Hazcam 3-D image from sol 490 shows the rover's front wheels nuzzled up to the Larry's Lookout outcrop, near the summit of Husband Hill. The team wanted to reach out with the rover's arm to study the chemistry of the rocks at the top of the outcrop, but the rocks were perched above a layer of soft sand, making the approach difficult. There was also concern that small landslides, like the one in this image, might injure the rover. After several approaches, however, the rover was able to reach the outcrop.

Above: This is a *Spirit* Pancam sol 500 color mosaic of the Larry's Lookout outcrop. From some angles, the profile of the outcrop resembled the front of a whale, and so this view became known as the "Whale Panorama." Many of these rocks showed signs of having been chemically changed, most likely by water.

entinels

Opposite: This *Spirit* Pancam image from sol 506 shows numerous details of the shape and texture of three large, isolated rocks encountered on the flanks Husband Hill, near the summit. rocks are pitted, perhaps in t from billions of years of dblasting by the wind, and haps in part because they may volcanic rocks, erupted from ery gassy, bubbly lava flow. middle rock is about 40 cm inches) across.

Above: This sol 506 *Spirit* Pancam image shows the same rocks in color. The top of the middle rock is very dusty, suggesting that the rocks have been in this static condition for a long time. Details can be seen even in the shadowed parts of these rocks; in general, the Martian sky is so bright that it provides enough light to study the geology of even shadowed regions of a surface.

Summit

↩ **Opposite:** Finally, after more than 400 sols of climbing, *Spirit* reached the summit of Husband Hill, roughly 100 meters (330 feet) above the plains where the rover landed. This Navcam 3-D mosaic from the summit was taken on sol 582. The rover spent more than 100 sols exploring the rocks, soils, and outcrops along this broad summit plateau.

Above: This *Spirit* Pancam panorama was taken on sol 582 from the rover's perch at the summit of Husband Hill, looking south toward Tennessee Valley and the Southern Basin of the Columbia Hills. As this and other similar views showed, the summit region is a windy place, and abundant evidence of sand dunes and wind-carved rock faces are visible in high-resolution Pancam images. On several occasions, the winds at the summit cleaned dust off the rover's solar panels (foreground), producing large surges in the rover's solar power supply.

Ventifacts

↻ **Opposite:** This *Spirit* Pancam 3-D image shows a typical wind-carved rock surface at the summit of Husband Hill. Fine lines and straight gouges carved into rocky surfaces by the wind are called "ventifacts," and are common geologic features in windy envi-ronments on both Earth and Mars. The part of the rock shown here is about 60 cm (24 inches) across.

Above: The same Husband Hill summit rock seen in the 3-D image opposite is shown here in this Pancam sol 623 color image. Color views like this show that many of the rocks in Gusev Crater are covered or coated by dust, which falls continually out of the atmosphere from the planet's famous dust storms. The expecta-tion that many rocks would have such dusty coatings is one reason why the rover can both brush and grind rock surfaces using the RAT.

Martian Sculpture

↻ **Opposite:** Sometimes an ancient surface can exhibit strange and exotic landforms. These rocks are good examples, photographed by *Spirit*'s Pancam on sol 692 during the descent from the summit of Husband Hill into the Southern Basin. The layered, elongated rock in the near field is about 50 cm (20 inches) long. The tall "L"-shaped rock behind that one (named "Tomahawk") is about 40 cm (16 inches) tall. These shapes resulted from wind sculpting, volcanic lava eruptions, impact cratering-processes, or some combination of these factors.

Above: This *Spirit* Pancam sol 692 image shows in false colors the same rocks seen opposite; the colors were chosen to accentuate the "dustiness" of the rock surfaces. Both Tomahawk and its wider nearby companion are relatively clean, indicating that they may be pieces of "ejecta" that arrived in the Columbia Hills only relatively recently. Alternatively, the rocks may have been continu-ally sandblasted free of dust by this windy environment.

El Dorado

↺ **Opposite:** Mars is covered in sand, and here's what it looks like up close, in this *Spirit* Microscopic Imager sol 709 3-D photo from the "El Dorado" sand dune south of the summit of Husband Hill. The donut-shaped indentation mark seen here was made by pushing a 3 cm (1.2 inch)–wide contact plate on one of the rover arm's spectroscopy instruments into the sand.

Above: This *Spirit* Pancam mosaic from sol 708 shows a broader view of the El Dorado sand dune field. The dunes cover an area about 150 meters (490 feet) across, along the southern flanks of Husband Hill. Rover measurements show that the sand is made up mostly of ground-down pieces of dark volcanic rock (basalt), unlike sand on Earth, which is made mostly of quartz. The relative lack of dust on the sand implies that the dunes may yet be moving, albeit at a very slow rate.

Gong Gong

↶ **Opposite:** This *Spirit* Pancam sol 736 3-D photo shows one of the most dramatic examples of rock texture ever seen on Mars. This small piece of rock, about 17 cm (7 inches) long and called "Gong Gong," is a heavily eroded piece of what geologists call scoria: a pitted, vesicle-covered, wind-carved volcanic rock. There must have been a high concentration of gas in this lava, judging by all the frothy bubbles that were formed when it erupted.

Above: This *Spirit* Pancam sol 736 color photo (above left) shows that what is left of the heavily eroded surface of Gong Gong is mostly dark, clean volcanic rock (basalt). However, the numerous small pits and mini "caverns" in the rock have trapped dust in some places. The spectacular microstructure of the rock is beautifully seen in the Microscopic Imager view (above right), which shows an area only 3 cm (1.2 inches) across.

Home Plate

↺ **Opposite:** After many months of descending the southern flanks of Husband Hill and working its way into the Southern Basin, *Spirit* finally arrived at the bright, circular feature named Home Plate. This sol 748 Pancam 3-D mosaic shows the edge of Home Plate, which turned out to be made of finely layered stacks of rocks, about 1.5 meters (5 feet) thick. The layers could be the result of sedimentary, volcanic, or even impact cratering processes.

Above: This *Spirit* Pancam sol 751 false color view shows a close-up of some of the fine layers along the edge of Home Plate. Individual layers are a few centimeters (about one inch) thick here. One of the layers appears to sag under the weight of a small embedded rock. This kind of feature is often called a "bomb sag" by geologists, and may result from a hot volcanic fragment erupting onto soft, perhaps even wet, volcanic ash deposits.

Salty Soil

↺ **Opposite:** The most trouble-some sign of "old age" on the *Spirit* rover by the time it reached Home Plate was the failure of the right front wheel motor. The team had to drive the rover backward, dragging that stuck wheel through the dirt. Luckily, dragging the wheel created a 15 cm (6 inch)– wide trench and dug up some bright, buried soils that we never would have seen otherwise, depicted in this sol 788 Pancam stereo view.

Above: This Pancam color view shows that there are several shades of bright soil dug up by the stuck wheel. Chemical analysis of the bright soils shows them to be made of concentrated, sulfur-rich salty minerals. It's still a mystery why these salty deposits are buried just below the surface near Home Plate. Sulfur-rich volcanic gases may be slowly percolating through and weathering the rocks, or they may be the evaporated remnants of salty liquid water.

Winter Haven

↺ **Opposite:** This *Spirit* Pancam 3-D mosaic is part of a large panorama taken on sols 814 to 932 from "Winter Haven," where the rover was parked in order to survive its second long, cold Martian winter. Because the Sun gets so low in the Martian sky in winter, there wasn't enough power to drive routinely, so the rover had to park on a hillside, tilting its solar panels into the weak winter sunlight just to stay alive.

Above: This is a Pancam color view of part of the large panorama taken from Winter Haven, called the "McMurdo" panorama. The long trench leading from the rover back toward the horizon was created by dragging the rover's stuck right front wheel through this soft dirt. Bright, sulfur-rich and silica-rich soils were exposed all along the way by this accidental wheel trenching. The rover remained parked at this location for over six months.

North ←

Endurance Crater
(sol 100 to 317)

Vostok Crater

Sol 350

Sol 400

Sol 100

Naturaliste
Crater

Voyager Crater

"Fram"

Sol 75

"Purgatory"

Sol 50

Landing Site: Eagle Crater
sol 1: Jan. 24, 2004

1 Kilometer

1 miles

OPPORTUNITY
Landed January 24, 2004
Meridiani Planum

Victoria Crater
(sols 952 to 1400+)

Sol 950

Sol 450

Erebus Crater
(sols 591 to 761)

Sol 900

Sol 800

Sol 500

Eagle Crater

↺ **Opposite:** The *Opportunity* lander bounced across the flat plains of Meridiani Planum and came to rest here in the middle of Eagle Crater, a small, shallow 20 meter (66 foot)–wide, 1.5 meter (4.9 foot)–deep impact crater. The rover spent nearly two months driving inside the crater and studying the fascinating outcrop rocks seen poking out of the crater's far inside wall. This *Opportunity* Pancam 3-D mosaic was shot on sols 58 to 60.

Above: This Pancam color mosaic of Eagle Crater was also shot on sols 58 to 60. The Eagle Crater landing site is dramatically different from the Gusev Crater landing site. For example, the surface has a dark, more brownish color than Gusev's redder, dustier soils. The small amounts of dust visible here occur mostly in small windblown drifts, or in a bright dusty "tail" around the east (right side) edge of the crater.

Blueberry Fields

↻ Opposite: One of the most stunning and surprising discoveries made by the *Opportunity* rover is that the surface of Meridiani Planum is covered with countless millions of small, spherical rock grains, like those seen here in this Microscopic Imager 3-D view from sol 19. The field of view here is about 3 cm (1.2 inches), and the spherical grains range from about 2 to 4 mm (0.08 to 0.16 inches) in diameter.

Above: In color images of these grains, team members noticed that they are grayer and less red than the dark, reddish-brown soils they rest on. This color difference is dramatically enhanced in false-color images, where the grains, dubbed "blueberries," burst out into bluish hues. This high-resolution false-color Pancam image (above left) shows a typical blueberry-filled scene on the floor of Eagle Crater. Above right is a wider-angle false-color image looking out of the crater and across the plains.

Snout

↶ **Opposite:** This *Opportunity* Pancam sol 13 3-D mosaic focuses on a close-up of the northern end of the Eagle Crater outcrop, centered on a small cracked rock named "Snout." The outcrop is small: Snout is only about the size of a loaf of bread, and the total thickness of the layered rocks exposed in the wall of the crater is only about 20 to 30 cm (8 to 12 inches).

Right: The soils in Meridiani Planum are darker than those *Spirit* studied in Gusev Crater. This color mosaic of the eastern end of the Eagle Crater outcrop shows that the rocks have a dusty-looking, reddish color. Unlike at Gusev, however, the dusty color of these rocks is not simply a surface coating that can be brushed or ground off. These are fundamentally different rocks, composed not of hard, dark volcanic minerals but of soft, layered, sedimentary deposits that are reddish throughout.

Layers and Berries

↺ **Opposite:** There was obviously some connection between the blueberries and the outcrop. This sol 15 stereo image of a region of blueberry-filled layered outcrop rock in Eagle Crater helped solve the mystery. The berries occur naturally inside the layers of the bright, reddish outcrop rocks. The bright rock is easy for the wind or other forces to break apart and erode away, allowing the berries to pop out of the rock and collect on the surface.

Above: Here, a special combination of Pancam's ultraviolet, visible, and infrared filters was used to construct a false-color "berry finder" map. Blueberries scattered across the ground, as well as those sitting on and embedded in the outcrop rock, light up as hot pink. Later imaging, chemical, and mineral measurements taken of the blueberries showed them to be rich in iron, probably formed in the same way that some minerals grow during the slow evaporation of liquid water.

Eagle Trench

↻ **Opposite:** The team needed to discover what lay below the blueberries and soil. The rovers don't have digging or drilling tools, but they can make shallow trenches in the soil using their wheels. This trench, excavated into the sandy soils of Eagle Crater on sol 26, is about 15 cm (6 inches) wide and 10 cm (4 inches) deep. The degree of layering of the subsurface soils and the sharpness of the trench walls provide unique information about the physical properties of the Meridiani soils.

Above: Again using the Pancam's false-color filters, trenches like this can be investigated to see if a layer of blueberries extends underground. Looking at the trench walls in this sol 26 mosaic, it's hard to see any evidence of blueberries more than a few centimeters (an inch or so) under the surface. Such pictures helped to show that the blueberries make up just a thin layer on the uppermost surface of the soils in Meridiani Planum.

Ratted Berries

↶ **Opposite:** *Opportunity*'s Rock Abrasion Tool (RAT) has been essential for understanding the history of this enigmatic landing site. On sol 30, the RAT was used to grind a 5 mm (0.2 inch)–deep hole into the Eagle Crater outcrop rock "McKittrick." The RAT ground down into two blueberries (lower middle, lower right), exposing their interiors in cross-section. None of the berries "shaved" this way showed evidence of internal layering or external coatings, further supporting the idea that they had been slowly "grown" from the evaporation of ancient, salty water.

Above: This sol 36 *Opportunity* Pancam false color image of the McKittrick RAT hole shows that the fine powdery rim outside the hole, which is made up partly of ground-up blueberries, is brighter and redder than the unground outcrop rock. This was a clue that the berries contain iron minerals, which often get brighter and redder when they are ground up into smaller sizes.

Fram

↺ **Opposite:** *Opportunity* headed eastward across the plains, towards a larger impact crater named "Endurance." Along the way, signposts were established at a number of interesting plains features, such as fractures and small craters. This Pancam 3-D image from sol 85 shows one of the rover's waypoints, the 8 meter (26 foot)–wide impact crater named "Fram." Fram is smaller than Eagle Crater, and thus not as deep.

Above: Bright sedimentary outcrop rocks are relatively easy to spot against the much darker, sandy and blueberry-covered soils of Meridiani Planum. This Pancam sol 88 color mosaic shows outcrop rocks revealed in and ejected from 8 meter (26 foot)–wide Fram Crater. The rover drove to Fram's rim and made chemical and mineral measurements that confirmed that the outcrop rocks here are essentially of the same kinds of salty sedimentary rock seen at Eagle Crater.

Dangerous Driving

↻ **Opposite:** Endurance Crater, at 130 meters (about 426 feet) wide and 20 meters (66 feet) deep, had steep slopes that made for spectacular vistas but potentially dangerous driving conditions. One wrong turn or bad slip across a slick berry-covered rock could mean disaster if the rover went over the edge. This Navcam 3-D image was taken near the edge of the rim of Endurance Crater on sol 115.

Above: The beautifully layered rocks seen on Endurance's inner rim wall near the upper center of this view, named "Burns Cliff," were quickly identified as the most interesting and accessible geologic target for study. It would be impossible to navigate the rover over the steep (60 to 90 degree) slopes, like those seen on the left side of this view. Instead, the drivers focused on the more gentle (10 to 25 degree) slopes, seen on the distant rim on the right side of this view.

Embedded Berries

↺**Opposite:** During the descent into Endurance Crater, the team used the microscope extensively to study the textures of different layers inside the crater. This *Opportunity* Microscopic Imager stereo image of a 3 mm (0.1 inch)–wide iron-rich blueberry embedded within sedimentary outcrop layers was shot from about 2 meters (6.5 feet) inside Endurance Crater on sol 142. The layers are extremely fine—so much so that a geologist would refer to these as "laminations" rather than layers.

Above: This broader sol 140 Pancam color view of blueberries poking out of outcrop rocks within Endurance Crater shows small "wind tails" growing downwind of the berries, as the outcrop rock surrounding them slowly erodes away. This is one way that scientists can deduce the modern-day average wind direction in this region. These kinds of surfaces are also potential driving hazards, because the rover's wheels could slip downhill if some of the berries get loosened.

Swiss Cheese

↰ **Opposite:** This is a sol 152 *Opportunity* Microscopic Imager 3-D image. It shows part of a 4.5 cm (1.8 inch)–diameter RAT hole ground into a very soft sedimentary rock layer about 3 meters (10 feet) along the traverse down into Endurance Crater. This rocky layer is very porous. Several holes or sockets can be seen (most prominently at center right) where blueberries popped out during the RAT grinding.

Above: During parts of the descent into Endurance Crater, the texture of the layers over which the rover drove changed dramatically across very short distances. This Pancam false color image from sol 150 shows three RAT holes that were ground into areas only 5 to 20 cm (2 to 8 inches) away from each other. Characterizing the chemistry, mineralogy, and physical properties of the layers provided the information needed to try to reconstruct the complex geologic history of the crater, along with the pre-crater surface.

Flaky

⤺ Opposite: At several points inside Endurance Crater, *Opportunity* encountered what appeared to be thin, flaky mineral veins growing within rock fractures. The small fractures seen here in this Pancam stereo view are only about 20 cm (8 inches) long. These veins may have been the result of groundwater forming mineral-rich rinds once it evaporated. Or they could represent sand or other mineral grains that flowed down and filled the cracks from above and that were somehow cemented together into the more resistant veins.

Above: Pancam color data from sol 170 didn't reveal any dramatic differences in color between the flaky veins and the surrounding host rock, suggesting that the flakes and the host rock may be of essentially the same material. Rover arm chemical measurements on some of the flakes suggest that water-formed mineral rinds may be the best explanation for these features.

Upslope

↺**Opposite:** *Opportunity's*
Navcam took this dramatic stereo
mosaic on sol 171 while climbing
down Endurance Crater and tilted
downhill nearly 30 degrees. The
view near the center of the image
is toward Burns Cliff, a 2.5 meter
(8 foot)–high stack of spectacularly
layered rocks, about 80 meters
(260 feet) away from the rover.
To the right, the view is that seen
toward the rear of the rover,
looking up toward the steep slope
down which the vehicle
had been descending.

Above: This false-color
Opportunity Pancam mosaic from
sol 175 is an example of a typical
"side survey" imaging sequence
used to try to characterize the
textures and colors of layers
encountered by the rover during
the descent into Endurance
Crater. Several RAT holes ground
into some of the different layers
can be seen in the upper right
of the image.

Berry-Free RAT Hole

↻ **Opposite:** As the rover continued into Endurance Crater, the outcrop textures began to change. Close to 20 meters (66 feet) along the drive down, *Opportunity* used its RAT to grind into what turned out to be a very soft layer of outcrop rock called "Diamond Jenness." When the Microscopic Imager took this 3-D image of the resulting 7 mm (0.3 inch)–deep RAT hole, there were some interesting discoveries: no blueberries! Perhaps this layer was formed at a time before the berries had been formed in other rocks, or perhaps the berries had eroded away.

Above: This sol 180 Pancam color view of the Diamond Jenness RAT hole shows how carefully the rover drivers had to work. They had to position the rover and the arm instruments on the "cleanest" part of the outcrop layer to avoid letting any of the blueberries that had fallen from layers above contaminate the measurement. The RAT hole is about 4.5cm (1.8 inches) in diameter.

Field of Dunes

↩ **Opposite:** The enormous dune field at the bottom of Endurance Crater was a tantalizing but danger-ous target for *Opportunity*. We were as close as we'd ever been to a large collection of dark sand dunes—and we wanted to measure them in detail. However, as this sol 201 Pancam stereo image shows, the sand is a potential rover death trap. Some dunes were more than a meter (over 3 feet) tall. *Opportunity* would almost certainly get stuck. Fortunately, the rover was close to several dune tendrils, and so could study the material without actually going in.

Above: On Mars, giant craters like Endurance are also big sand traps. Over time, if a supply of sand remains available, more and more sand is trapped and these central crater dunes grow to essentially fill the entire bowl of the crater. Other craters encoun-tered by *Opportunity* have been completely flattened out this way.

Wopmay

↺ **Opposite:** Nearing the lowest point of the descent into Endurance Crater, *Opportunity* encountered a number of strange rocks with interesting and unique textures. One such example is the 1 meter (3.3 foot)–wide, 2 meter (6.6 foot)–long rock named "Wopmay," seen here in this Navcam sol 250 stereo image. Wopmay has a lumpy surface covered with polygon-shaped cracks, suggesting that its surface has perhaps been heavily weathered, or may even have undergone cycles of wetting and drying.

Above: This Pancam sol 251 color view of Wopmay shows some of the best examples of the lumpy textures and polygonal cracking seen on the surface of this rock. Wopmay could have fallen down from a layer higher up in the crater, or could be a block of ejecta that just happened to land here after the impact. Evidence that the rock originated from deeper layers comes from closer-up photos that show a relative lack of blueberries on its surface.

Racetrack

↺ **Opposite:** *Opportunity*'s ulti-
mate destination within Endurance
Crater was Burns Cliff, a stack of
spectacularly layered sedimentary
rocks. Approaching the cliff was
tricky. The rover drivers had to use
stereo images and results from
the test rover in California to plan
Opportunity's drive across a 20 to
30 degree slope. This Pancam
sol 287 stereo view approaching
Burns Cliff makes the terrain
seem relatively flat, until you realize that the rover and the cameras
are themselves tilted at more than
a 20-degree angle.

Above: This *Opportunity* Pancam
sol 287 color image shows addi-
tional details in the layering as
the rover approached Burns Cliff.
These layers are what geologists
call "crossbeds," and they are
most likely the lithified, exposed
interiors of what used to be sand
dunes on the Martian surface.
The way some of the crossbed
layers intersect one another is
typical of exposed ancient sand
dune crossbeds found on Earth.

Rub al Khali

↩**Opposite:** This is what the typical scenery in Meridiani Planum looks like. Meridiani is basically flat; the mini-dunes seen in this sol 125 Navcam stereo mosaic are only about 5 to 10 cm (2 to 4 inches) high from crest to trough. The heights, specific shapes, and spacings of these ripples all provide information on the strength, direction, and duration of the winds in this part of Mars.

Above: This *Opportunity* Pancam photo is part of the "Rub al Khali" 360-degree panorama acquired on sols 456 to 464 from the plains south of Endurance Crater. Rub al Khali means "empty quarter" in Arabic, and the name is certainly fitting for this stark and lonely landscape. This panorama was taken while the rover was stuck here, after plowing itself into a slightly taller sand ripple. It took six weeks of spinning the wheels backward to extract the rover.

Purgatory

↺ **Opposite:** This *Opportunity* Pancam sol 496 stereo image shows the 15 cm (6 inch)–wide, 25 cm (10 inch)–deep rut that was created when the rover got stuck in the soft sands north of Erebus Crater. The rover "plowed into" this little dune at the breakneck speed of about 4 cm per second (0.1 mph). To the rover, the dune wasn't really an obstacle: The wheels kept happily spinning, and there was no indication that anything was wrong.

Above: After spinning its wheels backward for six weeks, *Opportunity* finally got free of what had come to be known as "Purgatory Dune" because the team wasn't sure whether the rover would ever escape. This colorized Navcam sol 491 photo shows the old (faded) wheel tracks going into the ruts, and the newer (sharper) tracks coming out.

Heat Shield

↺ **Opposite:** During the landing in January 2004, *Opportunity* was protected from the fiery heat of atmospheric entry by a heat shield. Just before landing, the heat shield was jettisoned and crashed onto the surface far from the landing site. This *Opportunity* sol 335 Pancam stereo image shows the remains of the heat shield in high-resolution detail. The shield was originally conical but the impact appears to have flipped it inside out; the metallic foil layer of thermal blanketing seen here used to be the inner surface.

Above: The remains of the heat shield shining in the sunlight generated camera glints off the metallic foil surfaces in this sol 335 Pancam color view. The crash site is littered with mechanical debris like springs, bolts, and metallic shards.

Naturaliste

↻ **Opposite:** *Opportunity* encountered small impact craters on the drive south from Endurance Crater. Here, for example, is a sol 390 Pancam stereo mosaic of Naturaliste, a 15 meter (49 foot)–wide impact crater about 1 kilometer (0.6 miles) south of Endurance.

Above: High-resolution and color imaging of small craters like Naturaliste helped the team to map the occurrence of sedimentary outcrop rocks and the thickness of the overlying sand and blueberry layer throughout *Opportunity's* traverse in Meridiani Planum. The rocks here, while more than 3 kilometers (1.9 miles) from the landing site at Eagle Crater, have essentially the same water-formed, sedimentary chemistry as the Eagle Crater outcrops. Apparently, the watery environment that led to the formation of these rocks covered large parts of this region of the planet.

An Ocean of Sand

↻ **Opposite:** Sometimes driving *Opportunity* across the sandy plains of Meridiani Planum felt more like sailing a ship across a vast wavy sea. This Pancam sol 433 stereo mosaic looks south from the rover's position near Voyager Crater, about 1.8 kilometers (1.1 miles) south of Endurance Crater. The terrain is so flat here that the low sandy ripples are the only topographic features to be seen, all the way out to the horizon, which is more than 3 kilometers (approximately 2 miles) away.

Above: In many places, the sands of Meridiani Planum are organized into exquisite patterns. This Pancam southward-looking sol 447 color mosaic, for example, reveals interesting fine-scale details within low sandy ripples near Purgatory Dune. The wavy layering parallel to the ripple ridge near the center of the photo indicates that the ripples themselves are probably internally layered or sorted into different particle sizes.

Strawberry RAT Hole

↰ **Opposite:** Crossing the boundary from the darker plains north of Erebus Crater to the lighter plains to the south also saw an accompanying change in the texture and chemistry of the sedimentary outcrop rocks. Some of the rocks started showing evidence of thick rinds or coatings. For example, this *Opportunity* Microscopic Imager sol 558 stereo mosaic (inset) shows a 4.5 cm (1.8 inch)–diameter, 5 mm (0.2 inch)–deep RAT hole ground into a thick coating that occurs on top of soft outcrop rocks near the rim of Erebus Crater.

Above: This *Opportunity* Pancam sol 561 false-color image shows the accompanying RAT hole, as well as a second RAT brush spot, on the rock "Strawberry," located near Erebus Crater. The rover arm had to be carefully positioned in order to sample parts of the rind as well as the "normal" outcrop, so that differences in their respective chemistries could be determined.

Erebus Highway

↺ **Opposite:** As the rover crossed the boundary near Erebus Crater between the darker plains to the north and the lighter plains to the south, the layer of berries and sand got thinner and thinner until, in many places, outcrop rocks were widely exposed out on the plains. This *Opportunity* Pancam sol 592 stereo image shows a 1 meter (3.3 foot)–wide section of finely layered outcrop rocks exposed near the rim of Erebus Crater.

Above: For context, this *Opportunity* Pancam sol 593 color mosaic shows a typical view of mixed dark sand and bright outcrop near Erebus Crater. These outcrop patches were often referred to as the "Erebus Highway" by many on the team, because it was much easier to drive the rover over the hard, flat "pavement" of outcrop rocks than on the soft, sandy ripples.

Stacks

↩ Opposite: This view shows even more dramatic evidence of fine-scale layering of outcrop rocks near the rim of Erebus Crater. The team was able to take so many detailed images near the rim of Erebus Crater partly because *Opportunity* was experiencing problems with the shoulder joint motor on its arm; as a result, the engineers and rover drivers had to spend many days diagnosing and developing a work-around for the problem.

Above: Images using the Pancam's ultraviolet and infrared filters can provide unique information on subtle physical or compositional differences among outcrop rocks. Such images have to be generated as "false-color" views because they represent colors that are not visible to the human eye. Here a natural color (RGB) composite to the accompanying 3-D image (above left) shows that to the human eye, most of the rocks in this region would have roughly the same colors. However, using the ultraviolet and infrared filters to generate a false-color image (above right) reveals different shades of purple, blue, and red in this *Opportunity* Pancam sol 674 view.

Accidental Trench

↰ **Opposite:** The experience of getting stuck at Purgatory Dune resulted in the need to come up with new driving methods that would help the team avoid getting the rovers stuck. This turned out to be prudent planning, because around sol 836, the rover got stuck again in a soft, deeper-than-expected sand dune just north of Victoria Crater. This sol 836 Pancam stereo image shows the "accidental trench" dug by the wheels while the rover was slipping in the sand. Fortunately, it took only a few sols to get the rover out of this particular jam.

Above: An enhanced color composite Pancam photo taken on sol 836 of the accidental trench reveals that the wheels were able to dig underneath the thin surface layer of blueberries to reveal brighter, essentially berry-free soils below. The lag deposit of berries that was first discovered back in Eagle Crater appears to extend over most of *Opportunity*'s traverse north of Victoria Crater.

Cape St. Vincent

↰ **Opposite:** After more than two Earth years of driving south, getting stuck in the sand, and having a forced stop to repair a balky arm, *Opportunity* finally made it to Victoria Crater around sol 952. Victoria is the largest crater yet visited by either rover, at 800 meters (0.5 miles) across and 50 to 70 meters (165 to 230 feet) deep. Steep cliffs, 6 to 15 meters (20 to 50 feet) tall flank the crater's rim, exposing a variety of distinct rock layers. This sol 1121 Pancam stereo view of the 6 meter (20 foot)–high Cape St. Vincent promontory is typical of the spectacular new landscapes now being explored by the rover.

Above: Color imaging has been an important part of the reconnaissance of Victoria Crater. Color views like this Pancam sol 1121 mosaic provide a way to estimate the composition and physical properties of the rock layers remotely. Here the color has been enhanced to a false color image for a more dramatic version.

Driving Smarter, Driving Farther

↩ **Opposite:** Despite working wonderfully on Mars for more than four years, the rovers always have something new to learn. The rover team constantly is assessing the vehicles' performance, and occasionally the team can upload new software or generate new driving tricks to make the rovers "smarter." This *Opportunity* Navcam sol 1163 stereo image shows wheel tracks that were intentionally made "wiggly" so that the rover's automatic driving software would be better able to measure its progress.

Above: Other rover software improvements included better obstacle avoidance software. This Pancam sol 1162 color mosaic shows the results of a test of the software performed near the rim of Victoria Crater. The experience gained by the rovers on Mars and the team operating the rovers on Earth will hopefully help improve the operation of future rovers being designed and built to go to Mars, the Moon, and other places throughout the solar system.

Web Links and Other Resources

Web sites

Mars exploration rover mission information: http://marsrovers.jpl.nasa.gov

Spirit 3-D images: http://marsrovers.jpl.nasa.gov/gallery/3d/spirit

Opportunity 3-D images: http://marsrovers.jpl.nasa.gov/gallery/3d/opportunity

Pancam color imaging home page: http://pancam.astro.cornell.edu/

Details of Athena Mars Exploration Rovers' instruments and mission: http://athena.cornell.edu

The Planetary Society: http://www.planetary.org/

The Mars Society: http://www.marssociety.org/

Chronology of Mars exploration: http://nssdc.gsfc.nasa.gov/planetary/chronology_mars.html

Mars facts: http://seds.lpl.arizona.edu/nineplanets/nineplanets/mars.html

Books

Bell, Jim, ed. *The Martian Surface: Composition, Mineralogy, and Physical Properties*. Cambridge University Press; 2008.

Bell, Jim. *Postcards from Mars: The First Photographer on the Red Planet*. Dutton, 2006.

Carr, Michael H. *The Surface of Mars*. Cambridge University Press, 2006.

Croswell, Ken. *Magnificent Mars*. Free Press, 2003.

de Goursac, Olivier. *Visions of Mars*. Harry N. Abrams, 2005.

Hartmann, William K. *A Traveler's Guide to Mars*. Workman Publishing Co., 2003.

Raeburn, Paul, and Matthew Golombek. *Mars: Uncovering the Secrets of the Red Planet*. National Geographic Press, 2000.

Squyres, Steven. *Roving Mars: Spirit, Opportunity, and the Exploration of the Red Planet*. Hyperion, 2005.

Magazine Articles

Bell, Jim. "Have Brain, Must Travel," *Scientific American*, August 2007, 36-37.

_____ "Portraits from Mars," *Astronomy*, January 2007, 64-69.

_____ "The Red Planet's Watery Past," *Scientific American*, December 2006, 62-69.

_____ "Photographing Mars," *The Planetary Report*, November/December 2006, 12–18.

_____ "Backyard Astronomy from Mars," *Sky & Telescope*, August 2006, 40–44.

_____ "In Search of Martian Seas," *Sky & Telescope*, March 2005, 40–47.

_____ "Blazing a New Path," *Astronomy*, August 2003, 52–57.

_____ "Mineral Mysteries and Planetary Paradoxes," *Sky & Telescope*, December 2003, 34–40.

_____ "The Human Side of Mars Exploration," *The Planetary Report*, November/December 2003, 12-17.

Christensen, Philip R. "The Many Faces of Mars," *Scientific American*, July 2005, 32–39.

Petit, Charles W. "Making a Splash on Mars," *National Geographic*, July 2005.

INDEX

Note: **Bold** page numbers refer to captions and photographs.

Adirondack rock, **42–45**
Blueberry fields, 32, **94–95**, **98–105**, **108–111**, **116–117**, **130**, **140–141**
Bomb sag, **84–85**
Bonneville Crater, **viii–ix**, **36**, **46–47**, **50**
Boulder-covered plains, **50–51**
Burns Cliff, **106–107**, **114–115**, **122–123**
Cameras (eyes) of rovers. *See also* Machine vision
  dual, creating parallax and dimension, 4–5
  geological vision from, 2–4
  resolution of, 4
  stereo images (anaglyphs) from, 4–5
  types of, 4
Cape St. Vincent, **142–143**
Columbia Hills. *See also* Husband Hill
  Longhorn outcrop, **58–59**
  West Spur, **26–27**, **37**, **58–61**
Crossbeds, **122–123**
Daily routine, 21–28
Desolate Ridge, **8–9**
Dunes, 4–5
  bump in road, **16–17**
  field of, **118–119**
  prevalence of, **48–49**
  Purgatory, **90**, **126–127**
Dust
  coating rocks, **76–77**
  debilitating effects of, 16–20
  fineness (coarseness) of, **62–63**
  homogeneity of, **62**
  images illustrating, **40–41**
  particle size, **40**
Eagle Crater, **90**
  Eagle Trench, **100–101**
  Gusev Crater vs., **92**
  McKittrick outcrop, **102–103**
  Opportunity escaping, **14–15**
  Opportunity landing in, **92–93**
El Dorado, **37**, **80–81**
Endurance Crater, **90**
  Burns Cliff, **106–107**, **114–115**, **122–123**
  dune field in, **118–119**
  embedded blueberries at, **108–111**
  flaky mineral at, **112–113**
  lack of blueberries at, **116–117**
  RAT findings at, **110–111**, **114–117**
  steep downhill panorama, **114–115**
  steep slopes, dangerous driving at, **106–107**
  Wopmay in, **120–121**
Erebus Crater
  outcrop rocks near, **138–139**
  traverse path and, **91**
Erebus Highway, **136–137**
Fram Crater, **90**, **104–105**
Gong Gong, **82–83**

Gusev Crater, **36–37**. *See also* Columbia Hills; Husband Hill
  Adirondack rock, **42–45**
  Bonneville Crater, **viii–ix**, **36**, **46–47**, **50**
  described, 22
  dusty dunes, **48–49**
  Eagle Crater vs., **92**
  El Dorado dune field, **80–81**
  Gong Gong, **82–83**
  hollows, **52–53**
  Home Plate, **84–85**
  plains, **50–53**
  Pot of Gold, **54–55**
  soils, compared to Meridiani Planum, **96**
  Spirit landing in, 22
  Spirit traverse path, **36–37**
  Winter Haven (McMurdo panorama), **88–89**
  Wooly Patch, **56–57**
Hazcams, about, 4, 26
Heat shield, **128–129**
Home Plate, **84–85**
Husband Hill, **37**, **66–73**
  craggy ridge, **68–69**
  flanks of, **60–61**
  Larry's Lookout (Whale Panorama), **70–71**
  slippery sand on, **66–67**
  summit views, **74–77**
  textured, isolated rocks, **72–73**
  Thanksgiving Panorama, **64–65**
  Tomahawk viewed from, **78–79**
Landers
  of *Opportunity* , **92–93**, **128–129**
  of *Spirit*, **38–39**
  Spirit rolling off of, **2–3**
Landing sites, **2–3**, 21–22, **36**, **90**, **92–93**, **128–129**
Larry's Lookout (Whale Panorama) outcrop, **70–71**
Life, requisite conditions for, 29–34
Longhorn outcrop, **58–59**
Machine vision, 7–12
  applications of, 7–8
  defined, 7
  driving rovers and, 8–10
  mimicking human eye, 7
  robotic intelligence and, 8–12
Mars
  discoveries changing perception of, 29
  dust issues on, 16–20
  requirements for life and, 29–34
  temperature on, 13
  time, 21, 28
Mazatzal, **22–23**
McKittrick outcrop, **102–103**
Meridiani Planum. *See also* Blueberry fields; Endurance Crater; *Erebus references*

accidental trench, **140–141**
Blueberry fields, 32
Cape St. Vincent, **142–143**
described, 22
Eagle Trench, **100–101**
Fram Crater, **90, 104–105**
Naturaliste Crater, **90, 130–131**
Opportunity landing in, 22, **90, 92–93**
Opportunity traverse path, **90–91**
Purgatory Dune, **90, 126–127**
Rub al Khali, **124–125**
Snout outcrop, **96–97**
soils, compared to Gusev Crater, **96**
spherical rock grains discovered in,
   93–94
"Strawberry" RAT hole, **134–135**
Victoria Crater area, **91, 140–145**
Microscopic Imager (MI), about, vi, 2, 4
Naming formations, xi–xii
Naturaliste Crater, **90, 130–131**
Navcams, about, 4, 26
Opportunity. See also Endurance Crater;
Meridiani Planum; Rovers
   driving hazards at Endurance Crater,
      106–107
   escaping Eagle crater, **14–15**
   getting stuck, **124–127, 140**
   heat shield, **128–129**
   landing site, 22, **90**
   spherical rock grains discovered by, **93–94**
   tracks of, **14–15, 124–127, 144–145**
   traverse path, **90–91**
Opportunity images. *See also* Eagle Crater
Pancams, about, 4, 26
Pot of Gold, **54–55**
Purgatory Dune, **90, 126–127**
Radio signals, 8, 25–26
Robotic intelligence and exploration, 8–12
Rock Abrasion Tool (RAT)
   at Endurance Crater, **110–111, 114–117**
   function and description, 2
   revealing Mazatzal composition, **22–23**
   "Strawberry" hole, **134–135**
   using, photos and, 5
   work on Adirondack, **42–45**
   work on blueberries, **102–103, 110–111**
Rocky Hollows, viii, ix
Rovers. *See also Opportunity; Spirit*
   arms/fingers of, 2
   boulder obstacles, **32–33**
   charging batteries of, 14
   conceptualization and creation of, vi,
      1–2, 13–14
   conditions designed for, 13–14
   daily routine, 21–28
   decision-making ability of, 8–10
   "driving," 5, 8–10, **142–143**
   driving backwards over dune, **16–17**
   durability and productivity of, viii–x,
      19–20

eyes of. *See* Cameras (eyes) of rovers
   "feeling" their environment, 2
   getting stuck, **124–127, 140**
   height of, 2
   as human extensions, 1–2
   lacking dust-cleaning system, 19–20
   landing positions, 21–22, **36–37**
   "licensed" drivers of, 5
   most important daily act, 14
   potential "killers" of, 13–19–20
   purpose of, 1
   robotic intelligence and, 8–12
   stuck wheel, **86–87, 88**
   uplink process, 24–25
Rub al Khali, **124–125**
Salty soil/rock, 66, **86–87, 102–103, 104**
Scoria, **82–83**
Snout outcrop, **96–97**
Solar panels, xii, 14, 16–20, 30, **74, 88**
Sols, defined, vi
Spirit. See also Columbia Hills; Gusev
   Crater; Husband Hill; Rovers
   lander, **38–39**
   landing site, 22, **36**
   rolling off lander, **2–3**
   in slippery sand, **66–67**
   tracks of, **64–67**
   traverse path, **36–37**
"Strawberry" RAT hole, **134–135**
Temperature, on Mars, 13
Tennessee Valley, **74–75**
3-D Images. *See also specific images (bold*
   page numbers)
   expectations vs. results, viii–x
   scale and perspective, x–xi
   stereo images (anaglyphs) and, 4–5
Time, Mars and Earth, 21, 28
Tomahawk, **78–79**
Tracks
   of *Opportunity*, **14–15, 124–127, 144–145**
   of *Spirit*, **64–67**
Trench, accidental, **140–141**
Uplink process, 24–25
Victoria Crater area, **91, 140–145**
Volcanic materials, **viii–ix, 22–23**, 30, **40,
   44–45, 50–51, 68, 72–73, 80, 82–86**
Vostok Crater, **90**
Voyager Crater, **90, 132**
Water, Earth and, 30
Water, evidence of, 22, 29, 32, 34, **56–57,
   66, 70, 86, 98, 102–103, 112–113, 130**
West Spur, **26–27, 37, 58–61**
Whale Panorama, **70–71**
Winter Haven (McMurdo panorama), **88–89**
Wishbone, **30–31**
Wooly Patch, **56–57**
Wopmay, **120–121**